MONTE CARLO METHODS

VOLUME I

MONTE CARLO METHODS

VOLUME I: BASICS

Malvin H. Kalos

Paula A. Whitlock

Courant Institute of Mathematical Sciences
New York University

A Wiley-Interscience Publication

JOHN WILEY & SONS

New York • Chichester • Brisbane • Toronto • Singapore

Library of Congress Cataloging in Publication Data:

Kalos, Malvin H.
Monte Carlo methods.

"A Wiley-Interscience publication."
Includes index.
Contents: v. I. Basics.
1. Monte Carlo method. I. Whitlock, Paula A.
II. Title.
QA298.K35 1986 517.2′82 86-11009
ISBN 0-471-89839-2

Printed in the United States of America

10 9 8 7 6 5 4 3 2 1

PREFACE

This book has had a long gestation period. While it simmered on the back burner, we pursued research in various aspects of Monte Carlo methods and their application to the simulation of physical systems. Out of this diversity we believe we see a basic way of looking at the field.

It is unfortunate that some observers and even some specialists of Monte Carlo methods seem to regard Monte Carlo as a bag of miscellaneous devices. Often it comes across that way when applied. It is true that like many other technical endeavors, especially those with as intensely practical an outlook as Monte Carlo methods, a body of ingenious tricks has arisen, awaiting invocation as needed. But we believe—and hope that our book is successful in conveying both in a general and a practical sense—that there are a number of unifying ideas that underlie the study and use of good Monte Carlo methods.

The first is the importance of random walks—on the one hand as they occur in natural stochastic systems, and on the other, in their relation to integral and differential equations.

The other basic theme is that of variance reduction and, above all, of importance sampling as a technical means of achieving variance reduction. Importance sampling is the transformation of a basically straightforward random sampling method by changing variables or, what amounts to the same thing, by changing the underlying probability distribution while leaving a required mean unchanged. It is by no means the only method, nor in particular cases the best method, for variance reduction. But it offers a coherent point of view about variance reduction. In important cases it offers the theoretical possibility of zero variance. The use of approximations to variance minimizing transformations is a powerful technique for the introduction of a priori knowledge based on experience or approximate solution of the problem at hand into a still exact numerical treatment based on Monte Carlo methods.

We believe that these ideas have stood us well in our research in radiation transport, in statistical physics, and in quantum mechanics and have served to unify them intellectually. We offer them to our readers in the hope that our point of view will make the theory and practice of Monte Carlo more interesting and more effective.

This book is a distillation of some years of practice and thought about Monte Carlo methods. As such it has benefited from the ideas and suggestions of many friends and colleagues, too numerous to list in full. It would be remiss not to mention some of them, however, starting with Gerald Goertzel, who first introduced one of us (MHK) to the mixed joys of Monte Carlo on primitive computers, and to many of the basic ideas expressed in our book. Others from whom we have learned include particularly Harry Soodak, Eugene Troubetzkoy, Herbert Steinberg, Loup Verlet, Robert Coveyou, Phillip Mittleman, Herbert Goldstein, David Ceperley, Kevin Schmidt, and Geoffrey Chester. Notes of early lectures taken by Jacob Celnik were very helpful.

We gratefully acknowledge the help and encouragement of our many colleagues and students during the time this book was being written. We especially thank David Ceperley for giving the original lecture on which Chapter 5 was based. Youqin Zhong and John Halton gave numerous suggestions for improving earlier versions of the manuscript. We thank them for their efforts and hope the final book lives up to their expectations.

MALVIN H. KALOS
PAULA A. WHITLOCK

New York, New York
August 1986

CONTENTS

1. What Is Monte Carlo? **1**

1.1. Introduction / 1
1.2. Topics to Be Covered / 3
1.3. A Short History of Monte Carlo / 4

2. A Bit of Probability Theory **7**

2.1. Random Events / 7
2.2. Random Variables / 9
2.3. Continuous Random Variables / 15
2.4. Expectations of Continuous Random Variables / 18
2.5. Bivariate Continuous Random Distributions / 21
2.6. Sums of Random Variables: Monte Carlo Quadrature / 23
2.7. Distribution of the Mean of a Random Variable: A Fundamental Theorem / 25
2.8. Distribution of Sums of Independent Random Variables / 28
2.9. Monte Carlo Integration / 31
2.10. Monte Carlo Estimators / 35

3. Sampling Random Variables **39**

3.1. Transformations of Random Variables / 40
3.2. Numerical Transformation / 48
3.3. Sampling Discrete Distributions / 50
3.4. Composition of Random Variables / 53

3.5. Rejection Techniques / 61
3.6. Multivariate Distributions / 71
3.7. The $M(RT)^2$ Algorithm / 73
3.8. Application of $M(RT)^2$ / 83
3.9. Testing Sampling Methods / 86

4. Monte Carlo Evaluation of Finite-Dimensional Integrals **89**

4.1. Importance Sampling / 92
4.2. The Use of Expected Values to Reduce Variance / 103
4.3. Correlation Methods for Variance Reduction / 107
4.4. Antithetic Variates / 109
4.5. Stratification Methods / 112
4.6. General-Purpose Monte Carlo Integration Code / 115
4.7. Comparison of Monte Carlo Integration and Numerical Quadrature / 115

5. Statistical Physics **117**

5.1. Classical Systems / 117
5.2. Quantum Simulations / 123

6. Simulations of Stochastic Systems: Radiation Transport **129**

6.1. Radiation Transport as a Stochastic Process / 130
6.2. Characterization of the Source / 134
6.3. Tracing a Path / 136
6.4. Modeling Collision Events / 140

7. Random Walks and Integral Equations **145**

7.1. Random Walks / 145
7.2. The Boltzmann Equation / 148
7.3. Importance Sampling of Integral Equations / 149

8. Introduction to Green's Function Monte Carlo **157**

8.1. Monte Carlo Solution of Homogeneous Integral Equations / 158
8.2. The Schrödinger Equation in Integral Form / 160

8.3. Green's Functions from Random Walks / 162
8.4. The Importance Sampling Transformation / 165

Appendix 169

A.1. Major Classes of prn Generators / 170
A.2. Statistical Testing of prn Generators / 172
A.3. Pseudorandom Number Generation on Parallel Computers / 179

Index 183

MONTE CARLO METHODS

VOLUME I

1 WHAT IS MONTE CARLO?

1.1. INTRODUCTION

The name *Monte Carlo* was applied to a class of mathematical methods first by scientists working on the development of nuclear weapons in Los Alamos in the 1940s. The essence of the method is the invention of games of chance whose behavior and outcome can be used to study some interesting phenomena. While there is no essential link to computers, the effectiveness of numerical or simulated gambling as a serious scientific pursuit is enormously enhanced by the availability of modern digital computers.

It is interesting, and may strike some as remarkable, that carrying out games of chance or random sampling will produce anything worthwhile. Indeed some authors have claimed that Monte Carlo will never be a method of choice for other than rough estimates of numerical quantities.

Before asserting the contrary, we shall give a few examples of what we mean and do not mean by Monte Carlo calculations.

Consider a circle and its circumscribed square. The ratio of the area of the circle to the area of the square is $\pi/4$. It is plausible that if points were placed at random in the square, a fraction $\pi/4$ would also lie inside the circle. If that is true (and we shall prove later that in a certain sense it is), then one could measure $\pi/4$ by putting a round cake pan with diameter L inside a square cake pan with side L and collecting rain in both. It is also possible to program a computer to generate random pairs of cartesian coordinates to represent random points in the square and count the fraction that lie in the circle. This fraction as determined from many experiments should be close to $\pi/4$, and the fraction would be called an estimate for $\pi/4$. In 1,000,000 experiments it is very likely (95% chance) that the number of points inside the circle would range between 784,600 and 786,200, yielding estimates of $\pi/4$ that are between 0.7846 and 0.7862, compared with the true value of 0.785398 . . . .

The example illustrates that random sampling may be used to solve a mathematical problem, in this case, evaluation of a definite integral,

$$I = \int_0^1 \int_0^{\sqrt{1-x^2}} dx\, dy. \tag{1.1}$$

The answers obtained are statistical in nature and subject to the laws of chance. This aspect of Monte Carlo is a drawback, but not a fatal one since one can determine how accurate the answer is, and obtain a more accurate answer, if needed, by conducting more experiments. Sometimes, in spite of the random character of the answer, it is the most accurate answer that can be obtained for a given investment of computer time. The determination of the value of π can of course be done faster and more accurately by non–Monte Carlo methods. In many dimensions, however, Monte Carlo methods are often the only effective means of evaluating integrals.

A second and complementary example of a Monte Carlo calculation is one that S. Ulam[1] cited in his autobiography. Suppose one wished to estimate the chances of winning at solitaire, assuming the deck is perfectly shuffled before laying out the cards. Once we have chosen a particular strategy for placing one pile of cards on another, the problem is a straightforward one in elementary probability theory. It is also a very tedious one. It would not be difficult to program a computer to randomize lists representing the 52 cards of a deck, prepare lists representing the different piles, and then simulate the playing of the game to completion. Observation over many repetitions would lead to a Monte Carlo estimate of the chance of success. This method would in fact be the easiest way of making any such estimate. We can regard the computer gambling as a faithful simulation of the real random process, namely, the card shuffling.

Random numbers are used in many ways associated with computers nowadays. These include, for example, computer games and generation of synthetic data for testing. These are of course interesting, but not what we consider Monte Carlo, since they do not produce numerical results. A definition of a Monte Carlo method would be one that involves deliberate use of random numbers in a calculation that has the structure of a stochastic process. By *stochastic process* we mean a sequence of states whose evolution is determined by random events. In a computer, these are generated by random numbers.

A distinction is sometimes made between simulation and Monte Carlo. In this view, simulation is a rather direct transcription into computing terms of a natural stochastic process (as in the example of solitaire). Monte Carlo, by contrast, is the solution by probabilistic methods of nonprobabilistic problems (as in the example of π). The distinction is somewhat useful, but often impossible to maintain. The emission of radiation from atoms and its interaction with matter is an example of a natural stochastic process since each event is to some degree unpredictible (cf. Chapter 6). It lends itself very well to a rather straightforward stochastic simulation. But the average behavior of such radiation can also be described by mathematical equations whose numerical solution can be obtained using Monte Carlo methods. Indeed the same computer code can be viewed simultaneously as a "natural simulation" or as a solution of the equations by random sampling. As we shall also see, the latter point of view is essential in formulating efficient schemes. The main point we wish to stress here is that the same techniques yield directly both powerful and expressive simulation and powerful and efficient numerical methods for a wide class of problems.

We should like to return to the issue of whether Monte Carlo calculations are in fact worth carrying out. This can be answered in a very pragmatic way: many people do them and they have become an accepted part of scientific practice in many fields. The reasons do not always depend on pure computational economy. As in our solitaire example, convenience, ease, directness, and expressiveness of the method are important assets, increasingly so as pure computational power becomes cheaper. In addition, as asserted in discussing π, Monte Carlo methods are in fact computationally effective, compared with deterministic methods when treating many dimensional problems. That is partly why their use is so widespread in operations research, in radiation transport (where problems in up to seven dimensions must be dealt with), and especially in statistical physics and chemistry (where systems of hundreds or thousands of particles can now be treated quite routinely). An exciting development of the past few years is the use of Monte Carlo methods to evaluate path integrals associated with field theories as in quantum chromodynamics.

1.2. TOPICS TO BE COVERED

The organization of the book is into several major areas. The first topic addressed is a review of some simple probability ideas with emphasis on concepts central to Monte Carlo theory. For more rigorous information on probability theory, references to standard texts will be given. The next chapters deal with the crucial question of how random events (or reasonable facsimiles) are programmed on a computer. The techniques for sampling complicated distributions are necessary for applications and, equally important, serve as a basis for illustrating the concepts of probability theory that are used throughout.

Then we consider quadratures in finite-dimensional spaces. Attention is paid to the important and interesting case of singular integrands, especially those for which the variance of a straightforward estimate does not exist so that the usual central limit theorems do not apply. These are cases for which variance reduction methods have an immediate and direct payoff.

The final chapters deal with applications of Monte Carlo methods. An introduction and survey of current uses in statistical physics is given. The simulation of a simple example of radiation transport is developed, and this naturally leads to the solution of integral equations by Monte Carlo. The ideas are then used as a framework upon which to construct a relationship between random walks and integral equations and to introduce the fundamentals of variance reduction for simulation of random walks.

1.3. A SHORT HISTORY OF MONTE CARLO

Perhaps the earliest documented use of random sampling to find the solution to an integral is that of Comte de Buffon.[2] In 1777 he described the following experiment. A needle of length L is thrown at random onto a horizontal plane ruled with straight lines a distance d $(d > L)$ apart. What is the probability P that the needle will intersect one of these lines? Comte de Buffon performed the experiment of throwing the needle many times to determine P. He also carried out the mathematical analysis of the problem and showed that

$$P = \frac{2L}{\pi d}. \tag{1.2}$$

Some years later, Laplace[3] suggested that this idea could be used to evaluate π from throws of the needle. This is indeed a Monte Carlo determination of π; however, as in the first example of this chapter, the rate of convergence is slow. It is very much in the spirit of inverting a probabilistic result to get a stochastic computation. We would call it an *analog* computation nowdays.[4]

Lord Kelvin[5] appears to have used random sampling to aid in evaluating some time integrals of the kinetic energy that appear in the kinetic theory of gases. His random sampling consisted of drawing numbered pieces of paper from a bowl. He worried about the bias introduced by insufficient mixing of the papers and by static electricity. W. S. Gossett (as "Student"[6]) used similar random sampling to assist in his discovery of the distribution of the correlation coefficient.

Many advances were being made in probability theory and the theory of random walks that would be used in the foundations of Monte Carlo theory. For example, Courant, Friedrichs, and Lewy[7] showed the equivalence of the behavior of certain random walks to solutions of certain partial differential equations. In the 1930s Enrico Fermi made some numerical experiments that would now be called Monte Carlo calculations.* In studying the behavior of the newly discovered neutron, he carried out sampling experiments about how a neutral particle might be expected to interact with condensed matter. These led to substantial physical insight and to the more analytical theory of neutron diffusion and transport.

During the Second World War, the bringing together of such people as Von Neumann, Fermi, Ulam, and Metropolis and the beginnings of modern digital computers gave a strong impetus to the advancement of Monte Carlo. In the late 1940s and early 50s there was a surge of interest. Papers appeared that described the new method and how it could be used to solve problems in statistical mechanics, radiation transport, economic modeling, and other fields.[8] Unfortunately, the computers of the time were not really adequate to carry out more than pilot studies in many areas. The later growth of computer power made it possible to carry through more and more ambitious calculations and to learn from failures.

At the same time, theoretical advances and putting into practice powerful error-reduction methods meant that applications advanced far faster than implied by sheer computing speed and memory size. The two most influential developments of that kind were the improvements in methods for the transport equation, especially reliable methods of "importance sampling"[9] and the invention of the algorithm of Metropolis et al.[10] The resulting successes have borne out the optimistic expectations of the pioneers of the 1940s.

REFERENCES

1. S. Ulam, *Adventures of a Mathematician*, Charles Scribner's Sons, New York, 1976, pp. 196–197.
2. G. Comte de Buffon, Essai d'arithmétique morale, *Supplément à l'Histoire Naturelle*, Vol. 4, 1777.

*This information was communicated privately to MHK by E. Segre and by H. L. Anderson.

3. Marquis Pierre-Simon de Laplace, Theorie Analytique des Probabilités, Livre 2, pp. 365–366 contained in *Oeuvres Complétes de Laplace*, de L'Académie des Sciences, Paris, Vol. 7, part 2, 1886.
4. In *Geometrical Probability* by M. G. Kendall and P. A. P. Moran, Hafner Publishing Co., New York, 1963 are discussed Monte Carlo applications of Buffon's problems, pp. 70–73.
5. Lord Kelvin, Nineteenth century clouds over the dynamical theory of heat and light, *Phil. Mag*., series 6, **2**, 1, 1901.
6. Student, Probable error of a correlation coefficient, *Biometrika*, **6**, 302, 1908.
7. R. Courant, K. O. Friedrichs, and H. Lewy, On the partial difference equations of mathematical physics, *Math. Ann*., **100**, 32, 1928.
8. (a) N. Metropolis and S. Ulam, The Monte Carlo Method, *J. Amer. Stat. Assoc*., **44**, 335, 1949. (b) M. D. Donsker and M. Kac, The Monte Carlo Method and Its Applications, Proceedings, Seminar on Scientific Computation, November 1949, International Business Machines Corporation, New York, 1950, pp. 74–81. (c) A. S. Householder, G. E. Forsythe, and H. H. Germond, Eds., Monte Carlo Methods, NBS Applied Mathematics Series, Vol. 12, 6, 1951.
9. H. Kahn, Modifications of the Monte Carlo Method, Proceeding, Seminar on Scientific Computation, November 1949, International Business Machines Corporation, New York, 1950, pp. 20–27.
10. N. Metropolis, A. W. Rosenbluth, M. N. Rosenbluth, A. H. Teller, and E. Teller, Equations of state calculations by fast computing machines, *J. Chem. Phys*., **21**, 1087, 1953.

GENERAL REFERENCES ON MONTE CARLO AND SIMULATION

P. Bratley, *A Guide to Simulation*, Springer-Verlag, New York, 1983.

J. H. Halton, A retrospective and prospective survey of the Monte Carlo method, *SIAM Review*, **12**, 1, 1970.

J. M. Hammersley and D. C. Handscomb, *Monte Carlo Methods*, Methuen, London, 1964.

J. P. C. Kleijnen, *Statistical Techniques in Simulation*, Marcel Dekker, New York, 1974.

R. Y. Rubinstein, *Simulation and the Monte Carlo Method*, John Wiley and Sons, New York, 1981.

I. M. Sobol, *The Monte Carlo Method*, translated from the Russian by V. I. Kisin, Mir, Moscow, 1975.

Yu. A. Shreider, ed., *The Monte Carlo Method*, Pergamon, New York, 1966.

K. D. Tocher, *The Art of Simulation*, D. Van Nostrand, Princeton, New Jersey, 1963.

2 A BIT OF PROBABILITY THEORY

2.1. RANDOM EVENTS

As explained in Chapter 1, a Monte Carlo calculation is a numerical stochastic process; that is, it is a sequence of random events. While we shall not discuss the philosophical question of what random events[1] are, we shall assume that they do exist and that it is possible and useful to organize a computer program to produce effective equivalents of natural random events.

We must distinguish between elementary and composite events. Elementary events are those that we cannot (or do not choose to) analyze into still simpler events. Normally the result (head or tail) of flipping a coin or the result (1–6) of rolling a die are thought of as elementary events. In the case of a die, however, we might interest ourselves only in whether the number was even or odd, in which case there are two outcomes. Composite events are those defined from a number of elementary events. Examples include flipping a coin twice (with four outcomes, head–head, head–tail, tail–head, tail–tail). It is sometimes useful to talk of this pair as a single "event."

As far as one knows, random events occur in nature; for example, the physical outcome of the scattering of an electron by an atom cannot be predicted with certainty. It is difficult to analyze with an assurance which natural random events are "elementary," although we shall have occasion to simplify our models of physical processes by treating a scattering event as elementary, and on that basis build up composite events. The distinction between an elementary random event and others depends on one's state of knowledge and the depth of the analysis given to the problem. Thus, one important kind of event, "compound elastic scattering" of neutrons, is usefully analyzed into a sequence of three elementary random events. On the other hand, "simple elastic scattering" is most likely an elementary event; that is, it is not possible to distinguish more basic stages.

Given an elementary event with a countable set of random outcomes, $E_1, E_2, \ldots, E_n, \ldots$, there is associated with each possible outcome E_k a number called a *probability*, p_k, which can lie between 0 and 1,

$$0 \le p_k \le 1.$$

If the kth outcome never occurs, $p_k = 0$; if it is sure to occur, $p_k = 1$. Conversely, if $p_k = 0$, we say that the event almost surely does not occur; and if $p_k = 1$, the event almost surely occurs. Another notation for the probability of event E_k is

$$P(E_k) = p_k.$$

Some simple properties of the probability are the following:

1. $P\{E_i \text{ and/or } E_j\} \le p_i + p_j$.
2. E_i and E_j are said to be ***mutually exclusive events*** if and only if the occurrence of E_i implies that E_j does not occur ($E_i \Rightarrow \bar{E}_j$) and vice versa. If E_i and E_j are mutually exclusive,

$$P\{E_i \text{ and } E_j\} = 0,$$
$$P\{E_i \text{ or } E_j\} = p_i + p_j.$$

3. A whole class of events can be mutually exclusive for all i and j. When the class is exhaustive, that is, all *possible* events have been enumerated,

$$P\{\text{some } E_i\} = \sum_i p_i = 1.$$

In the following we consider a compound experiment consisting of just two elementary events. For clarity, we imagine the first to have outcomes $\{E_i\}$ with probability p_{1i} and the second to have outcomes $\{F_j\}$ and probabilities p_{2j}, respectively. Each of p_{1i} and p_{2j} obeys statements 1, 2, and 3 above. An outcome of such a composite event is a pair (E_i, F_j).

4. The probability of the specific outcome (E_i, F_j) is p_{ij}, called the *joint probability* for E_i and F_j.
5. If $p_{ij} = p_{1i} \cdot p_{2j}$, then events E_i and F_j are independent.
6. Suppose E_i and F_j are not independent; then the joint probability

can be written

$$p_{ij} = \left(\sum_k p_{ik}\right)\left[\frac{p_{ij}}{\sum_k p_{ik}}\right]$$
$$= p(i)\left[\frac{p_{ij}}{\sum_k p_{ik}}\right]. \tag{2.1}$$

$p(i)$ defines a new number called the marginal probability for event E_i, that is, the probability that E_i does in fact occur, whatever the second event may be. Therefore,

$$\sum_i p(i) = \sum_i \sum_k p_{ik} = 1 \quad \text{and} \quad p(i) = p_{1i}.$$

The same holds true for the second event, F_j.

7. The second factor of Eq. (2.1) is the conditional probability

$$p(j \mid i) = \frac{p_{ij}}{\sum_k p_{ik}}$$

and is the probability for event F_j occurring, given that event E_i has occurred. The probability for *some* F_j should be 1, and indeed

$$\sum_j p(j \mid i) = \sum_j \frac{p_{ij}}{\sum_k p_{ik}} = \frac{\sum_j p_{ij}}{\sum_k p_{ik}} = 1 \quad \text{for every } i.$$

All joint probabilities can be factored into a marginal distribution and a conditional probability. This scheme can be generalized to treat the joint occurrence of three or more elementary events.

2.2. RANDOM VARIABLES

In many cases the outcome of a random event can be mapped into a numerical value, but in some circumstances it cannot (the probability of an event is always defined, but the assigning of a number to each outcome of a class of random events may not be useful). For example, when a photon interacts with an atom the photon may be scattered or it may cause other changes to happen within the atom. There is no useful way to assign a numerical value to correspond to the alternative changes. In simulating a queue, an empty queue could be equated with 0 length, but the meaning of the empty queue is really logical, not numerical. It implies that some other course of action must be taken. In general with simulations on the computer, the outcome of a random choice is often a logical event; it may imply that a different branch of the program is to be pursued. In the following discussion, however, we shall assume that for every elementary outcome E_i, there is associated a real number x_i. Such numbers are called random variables.

The *expectation* of this random variable x, that is, the stochastic mean value, is defined as

$$E(x) = \sum_i p_i x_i .$$

It is common in physics to write this as $\langle x \rangle$, and we shall often use that notation.

Consider some real-valued function

$$g(x_i) = g_i ,$$

where the x_i correspond to a countable set of elementary events with probabilities p_i. If x_i is a random variable, then $g(x_i)$ is also a random variable. The expectation of $g(x)$ is defined as

$$E(g(x)) = \langle g(x) \rangle = \sum_i p_i g(x_i). \tag{2.2}$$

This may be illustrated by analyzing the flipping of a coin, assigning 1 to heads, 0 to tails and using two different functions g_1 and g_2.

Event	p_i	x_i	$g_1(x) = 1 + 3x$	$g_2(x) = \dfrac{1+3x}{1+x}$
E_1 heads	$\frac{1}{2}$	1	4	2
E_2 tails	$\frac{1}{2}$	0	1	1
		$\langle x \rangle = \frac{1}{2}$	$\langle g_1(x) \rangle = \frac{5}{2}$	$\langle g_2(x) \rangle = \frac{3}{2}$

From the definition of the expected value of a function, we have the property that

$$\langle \text{constant} \rangle = \text{constant}$$

and that for any constants λ_1, λ_2 and two functions g_1, g_2,

$$\langle \lambda_1 g_1(x) + \lambda_2 g_2(x) \rangle = \lambda_1 \langle g_1 \rangle + \lambda_2 \langle g_2 \rangle. \tag{2.3}$$

In the table above g_1 is a linear function of x, so that

$$\langle g_1(x) \rangle = g_1(\langle x \rangle).$$

This is not true for the nonlinear function $g_2(x)$.

An important application of expectation values is to the powers of x. The nth moment of x is defined as the expectation values of the nth power of x,

$$\begin{aligned} \langle x^n \rangle &= \sum_i p_i x_i^n, \\ \langle x \rangle &= \sum_i p_i x_i = \mu. \end{aligned} \tag{2.4}$$

μ is called the expected or mean value. Similarly,

$$\langle x^2 \rangle = \sum_i p_i x_i^2.$$

The central moments of x are given by

$$\langle g_n(x) \rangle = \langle (x - \mu)^n \rangle = \sum p_i (x_i - \langle x \rangle)^n.$$

The second central moment has particular significance,

$$\begin{aligned} \langle (x - \mu)^2 \rangle = \langle (x - \langle x \rangle)^2 \rangle &= \sum_i p_i (x_i - \mu)^2 \\ &= \sum_i p_i x_i^2 - \langle x \rangle^2 = \langle x^2 \rangle - \langle x \rangle^2, \end{aligned} \tag{2.5}$$

and is called the variance of x (var$\{x\}$). The square root of the variance is a measure of the dispersion of the random variable. It is called the *standard deviation* and sometimes the *standard error*. The variance of

$g(x)$ can be determined as

$$\begin{aligned}\operatorname{var}\{g(x)\} &= \langle (g(x) - \langle g(x) \rangle)^2 \rangle \\ &= \sum_i p_i g^2(x_i) - \left(\sum_i p_i g(x_i) \right)^2 \\ &= \langle g^2 \rangle - \langle g \rangle^2. \end{aligned} \tag{2.6}$$

Consider two real-valued functions, $g_1(x)$ and $g_2(x)$. They are both random variables, but they are not in general independent. Two random variables are said to be independent if they derive from independent events.

As we have seen in Eq. (2.3), the expectations of a linear combination is the linear combination of the expectations. This result does not require that $g_1(x)$ and $g_2(x)$ be independent. The effect of statistical dependence will be seen in the variance of a linear combination of g_1 and g_2,

$$\begin{aligned}\operatorname{var}\{\lambda_1 g_1(x) + \lambda_2 g_2(x)\} &= \lambda_1^2 \operatorname{var}\{g_1(x)\} + \lambda_2^2 \operatorname{var}\{g_2(x)\} \\ &\quad + 2[\lambda_1\lambda_2\langle g_1 g_2\rangle - \lambda_1\lambda_2\langle g_1\rangle\langle g_2\rangle]. \end{aligned} \tag{2.7}$$

Let x and y be random variables. The expectation of the product is

$$\langle xy \rangle = \sum_{i,j} p_{ij} x_i y_j. \tag{2.8}$$

If x and y are independent, $p_{ij} = p_{1i} p_{2j}$ and

$$\langle xy \rangle = \sum_i p_{1i} x_i \sum_j p_{2j} y_j = \langle x \rangle \langle y \rangle.$$

The expectation of the product is now the product of the expectations. Assuming independence in Eq. (2.7), the bracketed quantity would disappear and

$$\operatorname{var}\{\lambda_1 x + \lambda_2 y\} = \lambda_1^2 \operatorname{var}\{x\} + \lambda_2^2 \operatorname{var}\{y\}. \tag{2.9}$$

This naturally leads to the definition of a new quantity, the covariance, which is a measure of the degree of independence of the two random variables x and y:

$$\operatorname{cov}\{x, y\} = \langle xy \rangle - \langle x \rangle \langle y \rangle. \tag{2.10}$$

The covariance equals 0 when x and y are independent and

$$\mathrm{cov}\{x, x\} = \mathrm{var}\{x\}.$$

Note that zero covariance does not by itself imply independence of the random variables. The following simple example illustrates that even functional dependence can still yield a zero covariance. Let x be a random variable that may be $-1, 0$, or 1, and define $y = x^2$. Obviously,

$$\langle x \rangle = 0,$$
$$\langle xy \rangle = \langle x^3 \rangle = 0,$$

so $\mathrm{cov}\{xy\} = \langle xy \rangle - \langle x \rangle \langle y \rangle = 0$.

The covariance can have either a positive or negative value. Another quantity which is related to the covariance is the correlation coefficient

$$\rho(x, y) = \frac{\mathrm{cov}\{x, y\}}{[\mathrm{var}\{x\}\,\mathrm{var}\{y\}]^{1/2}} \tag{2.11}$$

and

$$-1 \le \rho \le 1.$$

Since the covariance can be positive or negative, the variance of a linear combination of two dependent random variables can be greater or less than the variance if the variables were independent [cf. Eq. (2.7)]. A Monte Carlo calculation can try to take advantage of negative correlation as a means of reducing the variance, as will be discussed in Chapter 4.

2.2.1. The Binomial Distribution

Consider two events E_0 and E_1 that are mutually exclusive and exhaustive:

$$\begin{aligned} P\{E_1\} &= p, & x &= 1, \\ P\{E_0\} &= 1 - p, & x &= 0. \end{aligned} \tag{2.12}$$

The expectation values for the real number x and its square become

$$E(x) = p \cdot 1 + (1 - p)0 = p,$$
$$E(x^2) = p,$$

and the variance is then

$$\mathrm{var}\{x\} = p - p^2 = p(1-p).$$

Suppose that N independent samples of these events are drawn. Each outcome is either 0 or 1, and we set X to be the sum of the N outcomes:

$$X = \sum_{l=1}^{N} x_l .$$

Now the probability that $X = n$ is the probability that n of the x_l were 1 and $N - n$ were 0. That is,

$$P\{X = n\} = \binom{N}{n} p^n (1-p)^{N-n}. \tag{2.13}$$

This is the ***binomial distribution.*** $\binom{N}{n}$ is the binomial coefficient, which counts the number of different ways in which the n E_1's and $N - n$ E_0's may occur. Now suppose we associate the outcome $X_n = n$ with event E_n,

$$\langle X_n \rangle = \sum_n n \binom{N}{n} p^n (1-p)^{N-n} = Np. \tag{2.14}$$

This may be verified by direct computation of the sum, by consideration of the algebraic form $u\, d(u+v)^N/du$ with $u = p$, $v = 1 - p$, or by noting that the expected value of X is the sum of the expected value of all the x_l. The variance of the X_n is easily determined; since the X_n are independent the result in Eq. (2.9) may be employed,

$$\langle (X - Np)^2 \rangle = \sum_{l=1}^{N} \mathrm{var}\{x_l\} = \sum_{l=1}^{N} p(1-p) = Np(1-p). \tag{2.15}$$

2.2.2. The Geometrical Distribution

Suppose we carry out a certain experiment repeatedly and independently where there are only two outcomes: failure or success. If the outcome is a failure, the experiment will be repeated; otherwise, we stop the procedure. Now the random variable x of interest is the number of experiments we have done until the success appears. It is obvious that

$$P\{x = n\} = q^{n-1} p, \qquad n = 1, 2, \ldots,$$

where q is the probability of failure in one experiment, p is the probability of success in one experiment, and $p+q=1$. The average number of experiments being carried out is

$$\langle x\rangle=\sum_{n=1}^{\infty} nq^{n-1}p=\frac{p}{(1-q)^2}=\frac{1}{p}.$$

The variance of x can be calculated as

$$\begin{aligned}\mathrm{var}\{x\}&=\langle x^2\rangle-\langle x\rangle^2\\&=\left(\frac{2}{p^2}-\frac{1}{p}\right)-\frac{1}{p^2}\\&=\frac{1}{p^2}-\frac{1}{p}.\end{aligned}$$

As an example, in particle transport problems, the number of collisions one particle makes follows this distribution if the medium is infinite, homogeneous, and if the relative probability of different outcomes is constant.

2.2.3. The Poisson Distribution

A random variable x is said to follow a Poisson distribution when

$$P\{x=n\}=\frac{\lambda^n}{n!}e^{-\lambda},\qquad n=0,1,\ldots,$$

where λ is a parameter of this distribution. It is easy to find that

$$\begin{aligned}\langle x\rangle&=\lambda,\\\mathrm{var}\{x\}&=\lambda.\end{aligned}$$

This distribution is fundamental in the theory of probability and stochastic processes. It is of great use in applications such as research into queuing service systems and similar problems.

2.3. CONTINUOUS RANDOM VARIABLES

In the previous discussions we have assumed that the random events belonged to a discrete, countable set. Probabilities can be associated with continuous variables as well, giving rise to distribution functions. Such distributions are present both in nature and in artificial stochastic processes. As an example, consider the scattering of a photon by an atom. The angle at which the photon is scattered has values that are continuous between 0° and 180° with some angular intervals occurring more often than others.

Given that x is a real, continuous random variable,

$$-\infty < x < \infty,$$

a distribution function (or *cumulative* distribution function) is defined as

$$F(x) = P\{\text{a random selection of } X \text{ gives a value less than } x\}. \quad (2.16)$$

Suppose $x_2 > x_1$. Then $x_2 > x \geq x_1$ and $x < x_1$ are mutually exclusive events and exhaustive for $x \leq x_2$. Thus

$$P\{x_2 > x \geq x_1\} + P\{x < x_1\} = P\{x_2 > x\} \quad (2.17)$$

and

$$1 \geq P\{x_2 > x \geq x_1\} = P\{x < x_2\} - P\{x < x_1\} \geq 0. \quad (2.18)$$

$F(x)$ is a nondecreasing function of its argument. We may thereby conclude that $F(-\infty) = 0$ and $F(\infty) = 1$.

The distribution function may have intervals on which it is differentiable; in these intervals the probability density function (pdf) may be defined as

$$f(x) = \frac{dF(x)}{dx} \geq 0. \quad (2.19)$$

If $F(x)$ is not continuous, at the discontinuities discrete values are singled out. For example, imagine that $F(x)$ is piecewise constant everywhere except at a countable number of places (Figure 2.1); the distribution now describes a discrete set of random variables. Formally we may use the Dirac *delta function* to write

$$f(x) = \sum_l \delta(x - x_l) \times p_l,$$

where p_l is the jump of the distribution function at x_l. This emphasizes the fact that $f(x)$ need not be bounded.

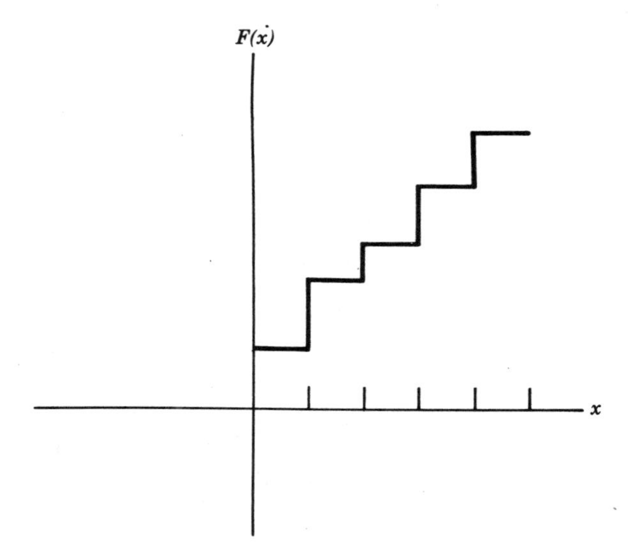

Figure 2.1. A piecewise-constant function.

Consider the distribution function shown in Figure 2.2. $F(x) = 0$ for all $x \leq 0$, so no x will be sampled here and the pdf $= 0$. In region II, $0 < x < 1$ and $F(x) = x$; the probability of sampling an x in the interval $0 \leq x_1 < x \leq x_2 \leq 1$ is $x_2 - x_1$. Put another way, since the pdf $= 1$, selecting a particular x is as likely as any other on (0, 1). For $x > 1$, $F(x) = 1$, and the probability of sampling an x in this region is 0.

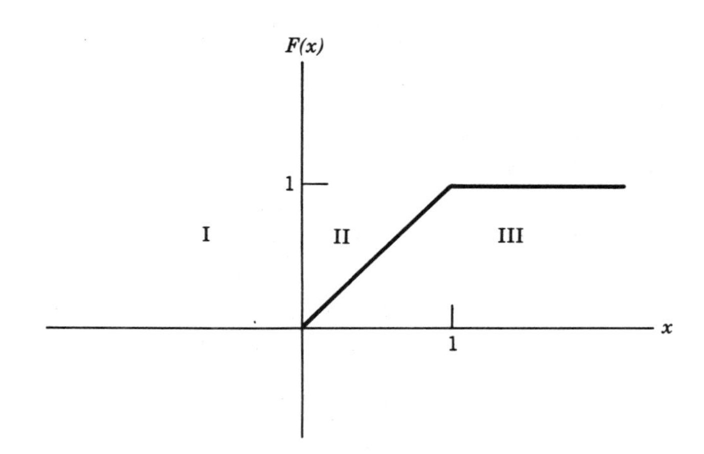

Figure 2.2. A discontinuous distribution function.

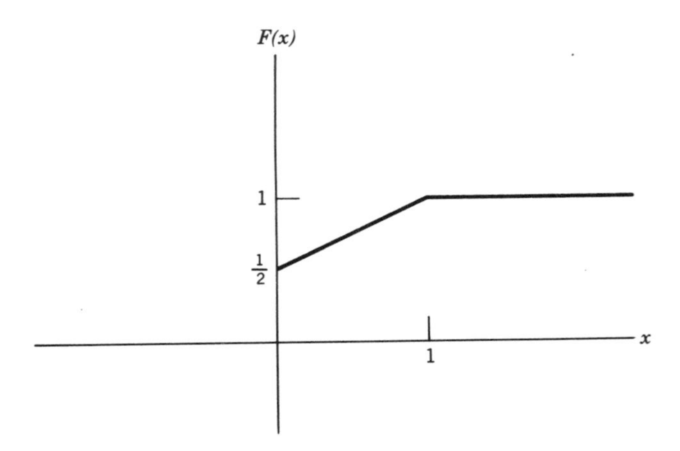

Figure 2.3. Discontinuous distribution function with a step discontinuity.

A slightly different example is the discontinuous distribution function in which F has a step discontinuity from 0 to $\frac{1}{2}$ at $x = 0$ shown in Figure 2.3. No matter how small an interval is chosen containing $x = 0$, the probability of finding an x in this interval is greater than or equal to $\frac{1}{2}$, so there is a finite probability for finding x exactly 0. For any other value of x, the pdf is continuous. We have a combination of a discrete and a continuous pdf. Such combinations occur in nature, as for example when an atom can undergo a radiative transition to either a continuum or a discrete level. The light spectrum, which may be considered as a pdf, will be a mixture of a continuous and a discrete part.

2.4. EXPECTATIONS OF CONTINUOUS RANDOM VARIABLES

The mean value of a continuous random variable is defined as

$$
\begin{aligned}
E(x) &= \int_{-\infty}^{\infty} x \, dF(x) \\
&= \int_{-\infty}^{\infty} x f(x) \, dx, \qquad (2.20)
\end{aligned}
$$

where $f(x)$ is the probability density function for x. The pdf has the normalization property

$$\int_{-\infty}^{\infty} f(x)\,dx = F(\infty) = 1.$$

The expectation value of any function of x is defined* similarly:

$$E(g(x)) = \int_{-\infty}^{\infty} g(x) f(x)\,dx,$$

and in particular

$$E(x^2) = \int_{-\infty}^{\infty} x^2 f(x)\,dx. \tag{2.21}$$

From Eqs. (2.20) and (2.21) the variance of x may be defined similarly to the discrete case

$$\text{var}\{x\} = E(x^2) - [E(x)]^2 = \langle x^2\rangle - \langle x\rangle^2. \tag{2.22}$$

The variance of a function of x becomes

$$\text{var}\{g(x)\} = E(g^2(x)) - [E(g(x))]^2. \tag{2.23}$$

In Table 2.1, $F(x)$, $f(x) = F'(x)$, the mean and variance are given for a few representative and interesting distributions. In particular the mean value of a random variable drawn from a normal distribution centered around μ is said to be μ. This result is easily proved by considering the following integral

$$\frac{1}{\sigma\sqrt{2\pi}} \int_{-\infty}^{\infty} (x-\mu) \exp\left[-\frac{(x-\mu)^2}{2\sigma^2}\right] dx = 0.$$

Since the integrand changes sign on reflection of x about μ, the integral may be rewritten

$$\frac{1}{\sigma\sqrt{2\pi}} \int_{-\infty}^{\infty} x \exp\left[-\frac{(x-\mu)^2}{2\sigma^2}\right] dx = \mu \frac{1}{\sigma\sqrt{2\pi}} \int_{-\infty}^{\infty} \exp\left[-\frac{(x-\mu)^2}{2\sigma^2}\right] dx.$$

*Since $g(x)$ is a random variable, it has its own pdf, and the expectation of $g(x)$ could be alternatively defined in the same way as in (2.20). It can be proved that both the definitions will be the same.[2]

TABLE 2.1. Continuous Probability Distribution Functions

Distribution Function	$F(x)$		$F'(x)$		$\langle x \rangle$	$\mathrm{var}(x)$
Uniform	$0,\ x<0$ $x,\ 0\le x\le a$ $1,\ x>a$	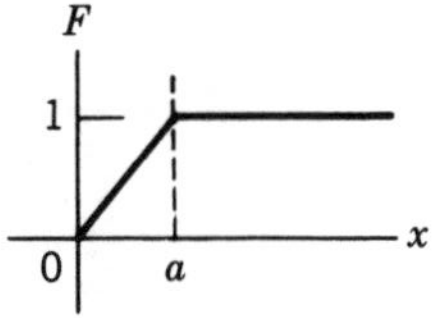	$0,\ x<0,\ x>a$ $\dfrac{1}{a};\ 0<x<a$	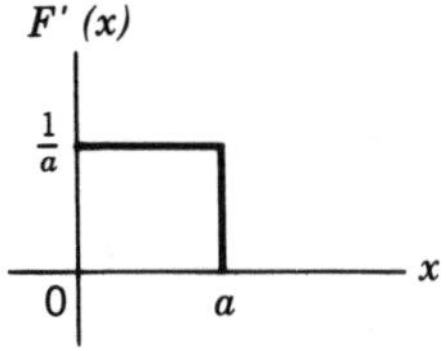	$\frac{1}{2}a$	$\frac{1}{12}a$
Exponential	$1-\exp(-\lambda x),\ x\ge 0$ $0,\quad x<0$	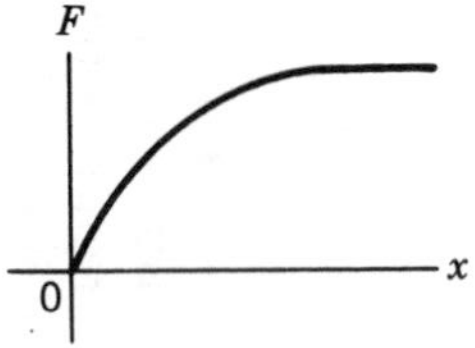	$0,\quad x<0$ $\lambda \exp(-\lambda x),\ x\ge 0$	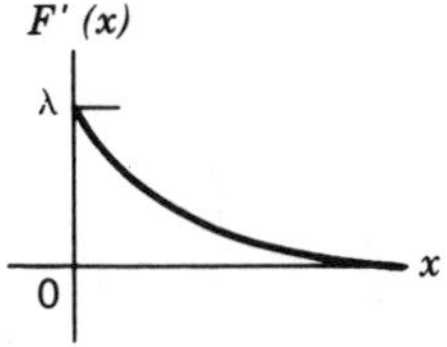	$\dfrac{1}{\lambda}$	$\dfrac{1}{\lambda^2}$
Normal $\phi(x\mid\mu,\sigma)$ μ = mean σ^2 = variance	$\dfrac{1}{\sigma\sqrt{2\pi}}\displaystyle\int_{-\infty}^{x}\exp\left\|\dfrac{-(t-\mu)^2}{2\sigma^2}\right\|dt$	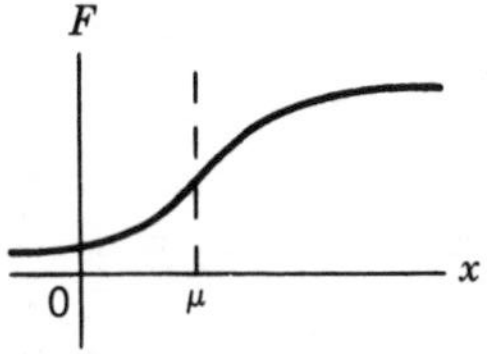	$\dfrac{1}{\sigma\sqrt{2\pi}}\exp\left\|-\dfrac{(x-\mu)^2}{2\sigma^2}\right\|dt$	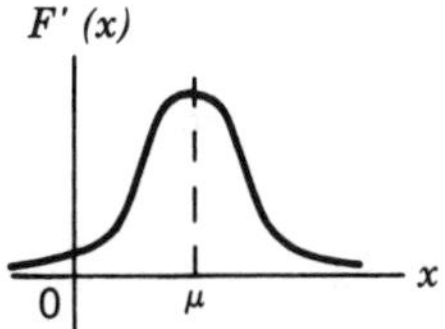 half width $\propto \sigma$	μ	σ^2
Cauchy (Lorentz)	$\dfrac{1}{2}+\dfrac{1}{\pi}\tan^{-1}\left(\dfrac{x}{a}\right)$	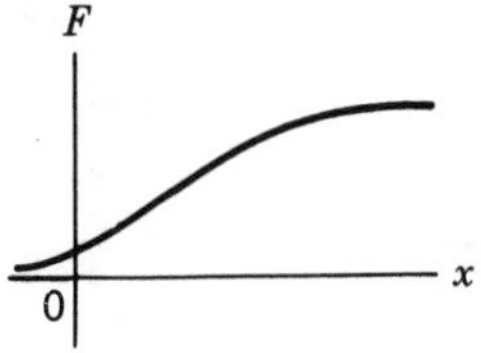	$\dfrac{a}{a^2+x^2}$	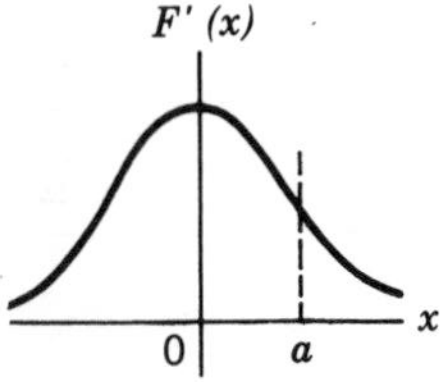	?	∞

The integral on the right-hand side is just the integral of the pdf; it equals 1, and the mean value of x is seen to be μ. The variance of the random variable is given by σ^2.

A less well-behaved example is provided by the Cauchy or Lorentz function. The mean value of a random variable sampled from a Cauchy distribution is

$$\frac{1}{\pi}\int_{-\infty}^{\infty} \frac{ax}{a^2 + x^2}\, dx.$$

To evaluate this improper integral using elementary calculus, the infinite endpoints of integration are replaced by finite quantities b and b', and the behavior of the integrand as b and b' approach infinity is considered. The integrand clearly diverges unless $b = b'$, and then the mean value is 0. This suggests that the mean value of a series of sampled random variables may be undefined unless the variables are chosen in some special way. The variance of a random variable sampled from a Cauchy distribution is infinity since the integral

$$\langle x^2 \rangle = \frac{1}{\pi}\int_{-\infty}^{\infty} \frac{ax^2}{a^2 + x^2}\, dx$$

diverges no matter how it is evaluated. In spite of the infinite variance, the Cauchy distribution can be sampled on the computer and used as needed.

Note that for each distribution there is a length scale, called variously a, λ^{-1}, or σ. As the scale becomes small, the normalization of the pdf grows large, inversely as the length scale, so as to ensure $\int f\, dx = 1$. For those distributions with a standard deviation σ, the width is proportional to σ.

2.5. BIVARIATE CONTINUOUS RANDOM DISTRIBUTIONS

A joint probability may be defined for continuous distributions

$$F(x, y) = P\{X \leq x,\ Y \leq y\}; \tag{2.24}$$

$F(x, y)$ is termed a *bivariate distribution*. The associated bivariate probability density function is

$$f(x, y)=\frac{\partial^2 F(x, y)}{\partial x\, \partial y} \tag{2.25}$$

and the expected value of any function of x, y is

$$E(g(x, y))=\langle g(x, y)\rangle=\int f(x, y) g(x, y)\, dx\, dy. \tag{2.26}$$

We define $\operatorname{cov}\{x, y\}$, $\rho(x, y)$ for continuous random variables x and y as was done in the discrete case, replacing sums by integrals. The variance has the following properties for any random variables x and y:

1. For a random variable c, which is constant (i.e., the random variable equals c with probability 1),

$$\operatorname{var}\{c\}=0.$$

2. For a constant c and random variable x,

$$\operatorname{var}\{c x\}=c^2 \operatorname{var}\{x\}.$$

3. For independent random variables x and y,

$$\operatorname{var}\{x+y\}=\operatorname{var}\{x\}+\operatorname{var}\{y\}.$$

If x and y are in general correlated, the joint pdf may be written

$$f(x, y)=\frac{f(x, y)}{\int f(x, y)\, dy} \int f(x, y)\, dy. \tag{2.27}$$

Let

$$m(x)=\int f(x, y)\, dy.$$

The integral $m(x)$ is called the marginal probability density function and by integrating over y, $m(x)$ has been reduced to a single distribution in x. The first factor in Eq. (2.27) is the conditional probability; that is, given an x, a y may be sampled from $f(y \mid x)=f(x, y) / \int f(x, y)\, dy$. The relationship in Eq. (2.27) is easily generalized to handle more than two correlated random variables. What happens to the marginal pdf and the conditional probability when x and y are independent [i.e., $f(x, y) = f(x)f(y)$] is left for the reader.

According to Eq. (2.27), for a given value of x, the random variable y has the conditional probability density function

$$f(y \mid x) = \frac{f(x, y)}{\int f(x, y)\, dy} = \frac{f(x, y)}{m(x)}.$$

Its expectation, called the *conditional expectation*, for fixed x is

$$E(y \mid x) = \int y f(y \mid x)\, dy = \frac{\int y f(x, y)\, dy}{\int f(x, y')\, dy'}$$
$$= \frac{\int y f(x, y)\, dy}{m(x)}.$$

The conditional expectation $E(y \mid x)$ is a function of the random variable x and is itself a random variable. The expectation of $E(y \mid x)$ is

$$E(E(y \mid x)) = \int E(y \mid x) m(x)\, dx.$$

Upon substituting in the definition for $E(y \mid x)$

$$E(E(y \mid x)) = \int \int y f(x, y)\, dy\, dx$$
$$= E(y).$$

A more general result is

$$E(E(g(x, y) \mid x)) = E(g(x, y)).$$

This result is very useful in the discussion of the "method of expected values" discussed in Section 4.2.

2.6. SUMS OF RANDOM VARIABLES: MONTE CARLO QUADRATURE

Suppose that the random variables $x_1, x_2, \ldots, x_n, \ldots$ are all drawn at random from the probability density function $f(x)$. A function G may be defined

$$G = \sum_{n=1}^{N} \lambda_n g_n(x_n), \tag{2.28}$$

where each of the g_n might be a different function of the x_n and λ_n is a real number. Each of the g_n is itself a random variable, and the sum of the $g_n(x_n)$ is a random variable. The expectation value of G becomes

$$\begin{aligned} E(G) = \langle G \rangle = E\Big(\sum_{n=1}^{N} \lambda_n g_n(x_n)\Big) \\ = \sum_{n=1}^{N} \lambda_n \langle g_n(x) \rangle \end{aligned} \tag{2.29}$$

since the expectation value is a linear operation. If all the x_n are independent, then the variance of G,

$$\mathrm{var}\{G\} = \langle G^2 \rangle - \langle G \rangle^2,$$

becomes

$$\mathrm{var}\{G\} = \sum \lambda_n^2 \, \mathrm{var}\{g_n(x)\}. \tag{2.30}$$

Let $\lambda_n = 1/N$ and all the $g_n(x)$ be identical and equal to $g(x)$; then the expectation value of G becomes

$$\begin{aligned} E(G) = E\Big(\frac{1}{N}\sum_{1}^{N} g(x_n)\Big) \\ = \frac{1}{N}\sum_{1}^{N} \langle g(x) \rangle = \langle g(x) \rangle \end{aligned} \tag{2.31}$$

or the function G, which is the arithmetic average of the $g(x)$, has the same mean as $g(x)$. G is said to be an "estimator" of $\langle g(x) \rangle$. More generally an expression G is an estimator of a quantity (such as $\int gf(x)\,dx$) if its mean $\langle G \rangle$ is a usable approximation of that quantity. The variance of G in equation (2.30) becomes

$$\begin{aligned} \mathrm{var}\{G\} = \mathrm{var}\Big\{\frac{1}{N}\sum_{1}^{N} g(x_n)\Big\} = \sum \frac{1}{N^2} \mathrm{var}\{g(x)\} \\ = \frac{1}{N} \mathrm{var}\{g(x)\}. \end{aligned} \tag{2.32}$$

That is, as N, the number of samples of x, increases, the variance of the mean value of G decreases as $1/N$. This result leads to the central idea of Monte Carlo evaluation of integrals; that is, an integral may be estimated by a sum

$$\langle g(x)\rangle = \int_{-\infty}^{\infty} g(x)f(x)\,dx = E\left(\frac{1}{N}\sum_{1}^{N} g(x_n)\right). \tag{2.33}$$

The method of using the relation given above is as follows: draw a series of random variables, x_n, from $f(x)$; evaluate $g(x)$ for each x_n. The arithmetic mean of all the values of g is an estimate of the integral, and the variance of this estimate decreases as the number of terms increases.

2.7. DISTRIBUTION OF THE MEAN OF A RANDOM VARIABLE: A FUNDAMENTAL THEOREM

In the discussion that follows on estimating integrals, it is assumed that the variance of the random variable always exists. If the variance does not exist, the mean value will converge, although more slowly. Alternatively, it will usually be possible to recast the sampling so that the variance does exist. A general method for doing this, along with several examples, will be given in Chapter 3.

The most general result of the kind we need is the "law of large numbers" of probability theory. Suppose the random variables $x_1, x_2, \ldots, x_N$ are independent and all drawn from the same distribution, so that the expectation of each x is μ. Then, as $N \to \infty$, the average value of the x's

$$\bar{x}_N = \frac{1}{N}\sum_{i=1}^{N} x_i$$

converges to μ almost surely:

$$P\{\lim_{N\to\infty} \bar{x}_N = \mu\} = 1.$$

There are stronger or weaker statements that can be made, but we shall not pursue them.[3]

The implication of the theorem is that the mean of n sampled random variables converges (in probability) to its expected value. In order to estimate the speed of convergence, we need stronger assumptions. The most important way of strengthening the hypothesis is to assume that the variance exists, which we do in the following.

Assume that an estimator G, its mean $\langle G\rangle$, and variance $\text{var}\{G\}$ all exist. Then the Chebychev inequality is

$$P\left\{|G-\langle G\rangle| \geq \left[\frac{\text{var}\{G\}}{\delta}\right]^{1/2}\right\} \leq \delta, \tag{2.34}$$

where δ is a positive number. This inequality could be called the first fundamental theorem of Monte Carlo for it gives an estimation of the chances of generating a large deviation in a Monte Carlo calculation. For definiteness let $\delta = \frac{1}{100}$. Then the inequality becomes

$$P\{(G-\langle G\rangle)^2 \geq 100\,\text{var}\{G\}\} \leq \frac{1}{100}$$

or, using (2.32) when $\text{var}\{G\} = (1/N)\,\text{var}\{g\}$,

$$P\left\{(G-\langle G\rangle)^2 \geq \frac{100}{N}\,\text{var}\{g\}\right\} \leq \frac{1}{100}.$$

Since by making N big enough, the variance of G becomes as small as one likes, the probability of getting a large deviation relative to δ between the estimate of the integral and the actual value becomes very small. For large sample size (large N), the range of values of G that will be observed with some fixed probability will be contained in a region of decreasing size near $\langle g\rangle$. This is the heart of the Monte Carlo method for evaluating integrals.

A much stronger statement than the Chebychev inequality about the range of values of G that can be observed is given by the central limit theorem of probability. For any fixed value of N, there is a pdf that describes the values of G that occur in the course of a Monte Carlo calculation. As $N \to \infty$, however, the central limit theorem shows that there is a specific limit distribution for the observed values of G, namely, the normal distribution (see Section 2.4). Set

$$G_N = \frac{1}{N}\sum_n g(x_n)$$

and

$$t_N = (G_N - \langle g \rangle)/[\mathrm{var}\{G_N\}]^{1/2}$$
$$= \frac{\sqrt{N}(G_N - \langle g \rangle)}{[\mathrm{var}\{g\}]^{1/2}};$$

then

$$\lim_{N\to\infty} P\{a \leq t_N < b\} = \int_a^b \frac{\exp[-t^2/2]\,dt}{\sqrt{2\pi}}. \tag{2.35}$$

Let $\sigma^2 = \mathrm{var}\{g\}$. Informally, the integrand of Eq. (2.35) can be rewritten

$$f(G_N) = \frac{1}{\sqrt{2\pi(\sigma^2/N)}} \exp\left[-\frac{N(G_N - \langle g \rangle)^2}{2\sigma^2}\right].$$

As $N \to \infty$, the observed G_N turns up in ever narrower intervals near $\langle g \rangle$ and one can predict the probability of deviations measured in units of σ.

The observed G_N is within one *standard error* (i.e., $\sigma/\sqrt{N}$) of $\langle g \rangle$ 68.3% of the time, within two standard errors of $\langle g \rangle$ 95.4% of the time, and within three standard errors 99.7% of the time.

The central limit theorem is very powerful in that it gives a specific distribution for the values of G_N, but it applies only asymptotically. How large N must be before the central limit theorem applies depends on the problem. If for a particular problem the third central moment μ_3 of g exists, then the central limit theorem will be substantially satisfied when

$$|\mu_3| \ll \sigma^3\sqrt{N}.$$

Without the central limit theorem, there is in general only the much weaker upper bound of the Chebychev inequality to suggest how much the observed G_N deviates from the actual mean. Of course in specific cases, studies can be made of the distribution of the estimator. Much Monte Carlo is done assuming that the theorem has been satisfied no matter what the sample size; reported errors must be considered optimistic in such cases.

When the variance is infinite it is sometimes possible to find a limit distribution for G that will lead to a central limit theorem for that particular problem. The limit distribution will in general not be normal. The Cauchy distribution yields an elementary example, as will be seen later.

The variance used in the discussion given above may itself be estimated using the observed independent values of $g(x_n)$ in the following way:

$$\left\langle \frac{1}{N}\sum_n g^2(x_n) - \left[\frac{1}{N}\sum_n g(x_n)\right]^2 \right\rangle = \langle g^2\rangle - \frac{1}{N^2}\left\langle \sum_n g_n^2 + \sum_n \sum_{\substack{m \\ n\neq m}} g_n g_m \right\rangle. \quad (2.36)$$

Using the independence of g_n and g_m in evaluating $\langle g_n g_m\rangle$, we find the right-hand side equal to

$$\left(1-\frac{1}{N}\right)\langle g^2\rangle - \frac{N(N-1)}{N^2}\langle g\rangle^2 = \frac{N-1}{N}\operatorname{var}\{g\}.$$

Thus an estimator for σ^2 is

$$\sigma^2 \cong \frac{N}{N-1}\left\{\frac{1}{N}\sum g^2(x_n) - \left(\frac{1}{N}\sum g(x_n)\right)^2\right\}. \quad (2.37)$$

A sample estimate of the variance of the estimated mean is given by

$$\operatorname{var}\{G_N\} \cong \frac{1}{N-1}\left\{\frac{1}{N}\sum g^2(x_n) - \left(\frac{1}{N}\sum g(x_n)\right)^2\right\}. \quad (2.38)$$

2.8. DISTRIBUTION OF SUMS OF INDEPENDENT RANDOM VARIABLES

Let x be sampled from $f_1(x)$ and y independently sampled from $f_2(y)$. If the sum $z = x + y$ is formed, what is the probability distribution function for z? The distribution function is defined as

$$F_3(z) = P\{x + y \leq z\}.$$

Since x and y are independent, the joint probability for x and y is

$$f(x, y) = f_1(x) f_2(y).$$

The variables x and y can be considered to form a point in the xy plane (see Figure 2.4). What fraction of the time does the point (x, y) lie

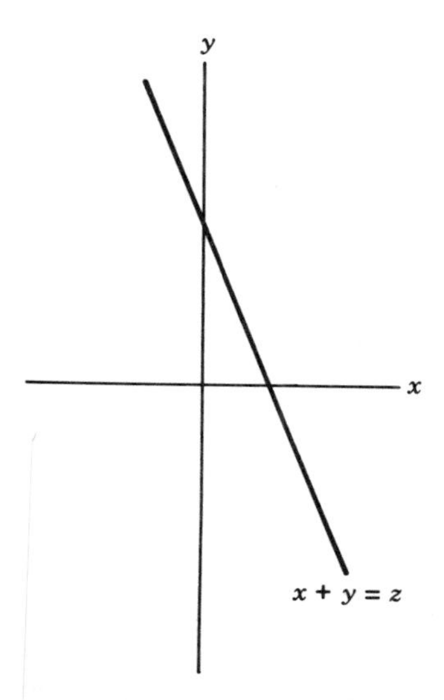

Figure 2.4. The sum of two random variables.

below the line $x + y = z$? The (cumulative) distribution function is

$$F_3(z) = \int_{x+y\le z}\int f_1(x) f_2(y)\, dx\, dy$$
$$= \int_{-\infty}^{\infty}\int_{-\infty}^{z-x} f_2(y) f_1(x)\, dy\, dx.$$

The cumulative distribution function of y is

$$F_2(y) = \int_{-\infty}^{Y} f_2(y)\, dy,$$

so

$$F_3(z) = \int_{-\infty}^{\infty} F_2(z - x) f_1(x)\, dx. \tag{2.39}$$

Differentiating with respect to z, one obtains for the pdf of z

$$f_3(z) = \int_{-\infty}^{\infty} f_2(z - x) f_1(x)\, dx;$$

this is a convolution, and Fourier transforms can be used to evaluate it. Let

$$c_1(t) = \int_{-\infty}^{\infty} e^{ixt} f_1(x)\, dx$$
$$= E(e^{ixt}) \tag{2.40}$$

In probability theory, $c_1(t)$ is labeled the characteristic function of x. The characteristic function of the sum $(x + y)$ is

$$c_3(t) = E(\exp[i(x + y)t])$$
$$= \int_{-\infty}^{\infty}\int_{-\infty}^{\infty} \exp[i(x + y)t] f_1(x) f_2(y)\, dx\, dy$$
$$= c_1(t)c_2(t), \tag{2.41}$$

or the characteristic function of a sum is the product of the characteristic functions of the terms of the sum.* Clearly induction gives the same result for a sum of n variables. The characteristic function may be inverted (by a Fourier transform) to give the pdf to which it corresponds. If n identical functions constitute the sum, this result can be used to prove the central limit theorem.

When $F(x)$ is the normal distribution (see Table 2.1), $\phi(x \mid 0, 1)$, the characteristic function is $\exp[-t^2/2]$. If a normal distribution is sampled n times, the sum is also distributed normally. The characteristic function is then $[c(t)]^n = \exp[-nt^2/2]$, which when inverted gives the normal distribution $\phi(x \mid 0, n^{1/2})$. A similar conclusion follows for the Cauchy distribution; the characteristic function is $\exp[-|t|]$, and after n samples, the characteristic function of the sum is $[c(t)]^n = \exp[-n|t|]$. The distribution of the sum of n Cauchy variables has a Cauchy distribution and the "width" of this Cauchy distribution increases as n. As a final example consider the exponential distribution $\lambda \exp[-\lambda t]$; its characteristic function is

$$c(t) = \frac{1}{1 - it/\lambda}.$$

Therefore a sum of exponential random variables will not be distributed exponentially.

*Note that if the two random variables are not independent, this statement is generally not true.

Note that the width of the distribution of the sum of n random variables increases with n. This is not contradictory with the earlier result on the mean of n random variables. It is not difficult to show—we shall prove it later—that if the characteristic distribution for x is $c(t)$, then the characteristic function x/n is $c(t/n)$. From this it follows that if n variables are drawn from the normal distribution $\phi(x \mid 0, 1)$, then the mean has characteristic function $\exp[-(n/2)(t/n)^2] = \exp[-t^2/2n]$. This may be inverted to give the distribution function $\phi(x \mid 0, n^{-1/2})$ for the mean. The latter shrinks with n as predicted by the central limit theorem. Indeed, we see that the limiting behavior holds exactly for any n.

In the same way the characteristic function for the mean of n Cauchy variables is $\exp[-n|t|/n] = \exp[-|t|]$. Thus the distribution of the mean is the same as for a single variable. Again the limit distribution for the mean is exact for any n, but now the distribution does not change its width.

2.9. MONTE CARLO INTEGRATION

We may summarize the important results of this chapter as follows. If $x_1, x_2, \ldots, x_n$ are independent random variables with probability density function $f(x)$ (x does not necessarily have to be in $\mathbb{R}^1$), then

$$G_N = \frac{1}{N}\sum_1^N g(x_n),$$

$$\langle G_N \rangle = \int f(x)g(x)\,dx,$$

and

$$\mathrm{var}\{G_N\} = \frac{1}{N}\mathrm{var}\{g\}.$$

As $N \to \infty$ and if the variance exists, the distribution of possible values of G_N narrows about the mean as $N^{-1/2}$; or the probability of finding a G_N some fixed distance away from $\langle G_N \rangle$ becomes smaller.

In the development of the Monte Carlo method so far, it has been assumed that the random variables are drawn from a continuous distribution function and that they are used to approximate an integral. Similar procedures can be employed to perform sums by Monte Carlo. It becomes advantageous to use Monte Carlo methods in the discrete case when many indices are involved. Consider the sum $\sum_i f_i g(x_i)$, where f_i is a discrete probability distribution. If random variables $x_1, \ldots, x_M$ are sampled from f_i and the quantity

$$G = \frac{1}{M} \sum_i^M g(x_i)$$

formed, the expectation value of G is an estimator for the sum

$$\langle G \rangle = \sum f_i g(x_i).$$

Monte Carlo evaluation of a sum might be used to determine the probability of winning at solitaire; the probability is a finite sum but contains a large number of terms. A sum over the permutations of L objects becomes cumbersome when L is large, but a Monte Carlo calculation can be performed efficiently.

The basic random variable used in Monte Carlo has been set by historical convention to be one distributed uniformly between 0 and 1. It was the first example presented in Table 2.1. The use of this random variable can be demonstrated in the following example. Consider the integral

$$\int_0^1 \sqrt{1-x^2}\, dx = \frac{\pi}{4}. \tag{2.42}$$

It can be rewritten

$$\int_0^1 f_u(x) \sqrt{1-x^2}\, dx = \frac{\pi}{4},$$

where

$$f_u(x) = 1, \qquad 0 \le x \le 1.$$

The integral is now in a form in which we can apply the method described above for evaluating integrals. A uniform random variable ξ_i is sampled from $f_u(x)$, $g(\xi_i) = \sqrt{1-\xi_i^2}$ is calculated and this process is repeated N times to form

$$G = \frac{1}{N} \sum g(\xi_i) = \frac{1}{N} \sum (1-\xi_i^2)^{1/2}. \tag{2.43}$$

Written in FORTRAN the process becomes

```
      SUM = 0.
      DO 10 I = 1,N
      SUM = SUM + SQRT(1. - RN(D)**2)
   10 CONTINUE
      MEAN = SUM/N.
```

We shall use RN(D) as a generic name for a utility for producing good pseudorandom numbers uniform on (0, 1). The evaluation of the integral in Eq. (2.42) by Monte Carlo can be approached in an entirely different manner.

Consider the unit square in the xy plane and the circle with unit radius (Figure 2.5). Integrating over the unit square but counting only those pairs of x and y that lie within the quarter circle yields the area of the quarter circle. That is

$$\int_0^1 \int_0^1 f(x, y) g(x, y)\, dx\, dy = \frac{\pi}{4},$$

where

$$f(x, y) = \begin{cases} 1 & (x, y) \text{ in } (0, 1) \otimes (0, 1) \text{ (i.e., inside unit square)} \\ 0 & \text{otherwise} \end{cases}$$

and

$$g(x, y) = \begin{cases} 1, & x^2 + y^2 \leq 1 \\ 0, & x^2 + y^2 > 1. \end{cases}$$

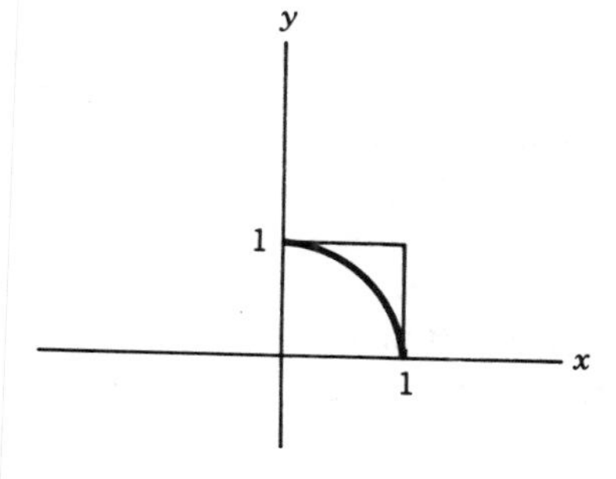

Figure 2.5. Unit square with inscribed quarter circle.

Since x and y are independent,

$$f(x, y) = f_u(x) f_u(y),$$

so that $f(x, y)$ may be sampled by drawing two independent uniform random variables. An algorithm for performing the Monte Carlo integration coded in FORTRAN is,

```
      SUM = 0.
      DO 10 I = 1,N
      X = RN(D)
      Y = RN(D)
      IF(X**2 + Y**2 .LE. 1.) SUM = SUM + 1.
   10 CONTINUE
      MEAN = SUM/N.
```

It is assumed that succeeding values of RN(D) are independent.

To evaluate the L-dimensional integral

$$\int \cdots \int g(x_1, x_2, \ldots, x_L)\, dx_1\, dx_2 \cdots dx_L,$$

L uniform random variables could be sampled, the function $g(x_1, x_2, \ldots, x_L)$ calculated, and the whole process repeated N times to generate an estimate for the integral.

When a variable is generated using RN(D) or a similar routine, it is a pseudorandom variable. Since it is possible to use real random variables in any calculation, why are pseudorandom variables used? An absolute requirement in debugging a computer code is the ability to repeat a particular run of the program. If real random numbers were used, an identical calculation could not be repeated and the recurrence of an error would be left to chance. It has been suggested that pseudorandom numbers be used to debug a program and that in the actual exercising of the program real random numbers be used. This method also suffers from the inability to repeat a particular calculation. If after many hours of computer time an error should occur in a code (subtle logical errors occur in many Monte Carlo codes), it is of utmost importance to be able to repeat the error at will as an aid to debugging. It is also very useful to be able to repeat a calculation when changes are made or when the program is moved to a different computer.

2.10. MONTE CARLO ESTIMATORS

We have defined an estimator as a useful approximation to a quantity of interest Q, which may be derived from a Monte Carlo calculation. In the example above Eqs. (2.42) and (2.43) in which

$$\frac{\pi}{4} = \int \sqrt{1-x^2}\,dx \cong \frac{1}{N}\sum_i \sqrt{1-\xi_i^2},$$

we consider Q to be $\pi/4$ and our estimator is the approximation on the right-hand side, which is a function $\theta(\xi_1, \xi_2, \ldots, \xi_N)$ of the N random numbers (or random variables) used in the calculation.

The function θ is of course itself random, and the statement that it gives a satisfactory approximation to Q means that it is not expected to fluctuate far from Q. Put a little more formally,

$$\langle(\theta - Q)^2\rangle / Q^2 \ll 1.$$

Acceptable values of the ratio depend on the application. We write

$$\langle(\theta - Q)^2\rangle = \langle(\theta - \langle\theta\rangle)^2\rangle + (\langle\theta\rangle - Q)^2$$

and observe that the quality of θ as a measure of Q comes separately from the variance of θ and from the departure of its mean from Q. The quantity $\langle\theta\rangle - Q$ is called the *bias* of the estimator. An unbiased estimator is one for which $\langle\theta\rangle = Q$, and the mean value of any experiment is just Q whatever the number N may be.

The quadratures we have discussed are unbiased since the result is linear in the functions calculated. For some problems, it is very difficult to formulate unbiased estimators. As we shall see, there are many problems for which the answer required is a ratio of integrals:

$$Q = \frac{\int_0^1 g_1(x)\,dx}{\int_0^1 g_2(x)\,dx},$$

for which a suitable estimator is

$$\theta(\xi_1, \ldots, \xi_N) = \frac{\sum_i g_1(\xi_i)}{\sum_i g_2(\xi_i)}.$$

Since this is not a linear function of the g_2, it is biased. An example that can easily be analyzed is

$$1 = \frac{1}{\int_0^\infty xe^{-x}\,dx} = \frac{N}{\sum_{i=1}^{N}(-\log(\xi_i))}.$$

The random variable x is sampled from an exponential distribution (see Section 3.1) and an estimator for the quotient is formed. It can then be shown that

$$\left\langle \frac{N}{\sum(-\log \xi_i)} \right\rangle = \frac{N}{N-1} \to 1 + \frac{1}{N} + \frac{1}{N^2} + \cdots.$$

Our estimator is biased by $1/N$, which, for large N, decreases faster than the standard error of the mean ($\sigma/\sqrt{N}$). This $1/N$ behavior is typical of the bias of such ratios. The results that may be derived from a Monte Carlo calculation are more general than this, and may have different variation of the bias. It is of course best if the bias becomes 0 as N grows large.

An estimator θ is termed *consistent* for the quantity Q if θ converges to Q with probability 1 as N approaches infinity. That is, θ is a consistent estimator of Q if

$$P\{\lim_{N\to\infty} \theta(x_1, x_2, \ldots, x_N) = Q\} = 1.$$

The law of large numbers is the statement that the sample mean $\bar{x}_N$ is a consistent (and unbiased) estimator of the mean μ. It further implies that estimators of quotients that are quotients of means are also consistent (although, in general, biased).

While unbiased estimators are desirable, they should not be introduced at the expense of a large variance, since the overall quality is a combination of both. In general one seeks the minimum of $\langle(\theta - Q)^2\rangle$.

An example may clarify the issue. Suppose $\{x_i\}$ are drawn uniformly and independently on the interval (0, 1) and we wish to estimate the mean Q from the sampled x_i. The obvious estimator is the mean

$$\bar{x}_N = \theta_1 = \frac{1}{N}\sum x_i.$$

A plausible alternative for large N is

$$\theta_2 = \tfrac{1}{2}\max(x_1, x_2, \ldots, x_N).$$

Note that this estimator always gives results less than the mean, $\frac{1}{2}$. It is easy to show that

$$E(\theta_1) = Q,$$

$$\mathrm{var}\{\theta_1\} = \frac{Q^2}{3N} = 0\left(\frac{1}{N}\right),$$

while

$$E(\theta_2) = \frac{N}{N+1}Q = Q\left(1 + 0\left(\frac{1}{N}\right)\right),$$

$$E(\theta_2 - Q)^2 = \mathrm{var}\{\theta_2\} = \frac{2Q^2}{(N+1)(N+2)} = 0\left(\frac{2}{N^2}\right).$$

Thus, although θ_2 is biased (by a fractional amount $1/N$), its variance is smaller by about a ratio of $N/6$. For some purposes, this would be a more useful estimator. Note, however, that the bias is larger than the standard error $\sqrt{2}/N$ by a constant ratio.

Just as a good Monte Carlo calculation must be supplemented with an estimate of the statistical error, sources of bias should be identified. The bias should be estimated numerically and corrected for or an upper bound should be determined. A useful way of estimating bias when the behavior with N is known is to use samples smaller than N, say $n = N/m$, and average the more biased estimator over the m groups obtained (bearing the previous estimator in mind):

$$\text{bias of } \frac{\sum_{i=1}^{N} g_1(\xi_i)}{\sum_{i=1}^{N} g_2(\xi_i)} \cong \frac{c}{N},$$

$$\text{bias of } \frac{1}{m}\sum_{l=1}^{m}\left(\frac{\sum g_1(\xi_i)}{\sum_{i=n(l-1)+1}^{nl} g_2(\xi_i)}\right)_{\text{group}_l} = \frac{c}{n} = \frac{cm}{N}.$$

We note in passing that this method of grouping is also a practical way of estimating the variance of the quotient.

REFERENCES

1. M. Kac, What is random?, *American Scientist*, **71**, 405, 1983; P. Kirschenmann, Concepts of randomness, *J. Philos. Logic*, **1**, 395, 1972.

2. J. L. Doob, *Stochastic Processes*, John Wiley and Sons, New York, 1953. Expectation values are discussed on pp. 12 and 13. Equality of the two definitions is proved on pp. 617–622.

3. P. Révész, *The Laws of Large Numbers*, Academic Press, New York, 1968.

GENERAL REFERENCES FOR FURTHER READING

Elementary

K. L. Chung, *Elementary Probability Theory with Stochastic Processes*, Springer-Verlag, New York, 1979.

H. Cramér, *The Elements of Probability Theory*, John Wiley and Sons, New York, 1958.

More Advanced

H. Cramér, *Random Variables and Probability Distributions*, Cambridge University Press, London, 1970.

W. Feller, *An Introduction to Probability Theory and Its Applications*, Vols. 1 and 2, John Wiley and Sons, New York, 1950.

3 SAMPLING RANDOM VARIABLES

We have sketched how a Monte Carlo calculation is a numerical stochastic process. The next step consists in designing and carrying out such a process so that answers to interesting questions may be obtained. In so doing, it is usually required that random variables be drawn from distribution functions that define the process. For example, to evaluate the integral $\int f(x)g(x)\,dx$, values of x must be drawn from $f(x)$ and the average value of $g(x)$ over a set of such x calculated. The process of "sampling x from $f(x)$," as it is ordinarily called, is therefore an essential technical matter. It is the purpose of this chapter to introduce the reader to the methods required. It will be beyond our scope to give a complete review or survey of methods or of known algorithms. References will be supplied to other literature.[1] Our treatment may serve to illustrate important principles, to exercise ideas of probability, but above all to demonstrate that sampling any $f(x)$ can in fact be carried out. At the same time, some specific techniques and results will be presented.

First, we must define what we mean by *sampling*. Consider some space Ω_0 and $x \in \Omega_0$, together with a probability density function* $f(x)$, where

$$\int_{\Omega_0} f(x)\,dx = 1.$$

A sampling procedure is an algorithm that can produce a sequence of values of x ("random variables") $x_1, x_2, \ldots$ such that for any $\Omega \in \Omega_0$,

$$P\{x_k \in \Omega\} = \int_{\Omega} f(x)\,dx \leq 1. \tag{3.1}$$

*The case in which the space is discrete and the case of mixed discrete and continuous variables need only a slight change of notation.

For a one-dimensional distribution defined on (0, 1), this means that

$$P\{x_k \in (a, b)\} = \int_a^b f(x)\, dx, \qquad 0 < a < b < 1.$$

Informally, for small values of $b - a = dx$,

$$P\{x_k \in dx\} = f(x)\, dx.$$

It will be possible to do this only by already having a sequence of some basic random variables. It has become conventional to start with random variables that are independent and uniformly distributed on (0, 1). We shall denote these $\xi_1, \xi_2, \ldots$, and assume that they can be generated at will by a computer procedure called RN(D). Such routines are widely available, usually giving satisfactory imitations of truly random variables. These are called pseudorandom numbers. It is important to note that here *satisfactory* means that the results are adequate in a particular context. No general method has ever been proved acceptable in any but the most elementary calculations, and well-known computer manufacturers have supplied bad pseudorandom generators. It is unfortunately necessary to test such generators both intrinsically and in the context of a specific class of applications. Further discussion of these issues will be given in the appendix.

3.1. TRANSFORMATIONS OF RANDOM VARIABLES

In the discussion that follows, an indefinite supply of uniform pseudorandom variables is assumed to exist. Suppose that x is a random variable with cumulative distribution function $F_X(x)$ and pdf

$$f_X(x) = \frac{dF_X}{dx}$$

and that $y = y(x)$ is a continuous nondecreasing function of x as in Figure 3.1. What is $F_Y(y)$? The variable x and the function $y(x)$ map into each other

$$y(X) \leq y(x) \quad \text{iff} \quad X \leq x, \tag{3.2}$$

so the probabilities become

$$P\{y(X) = Y \leq y(x)\} = P\{X \leq x\} \tag{3.3}$$

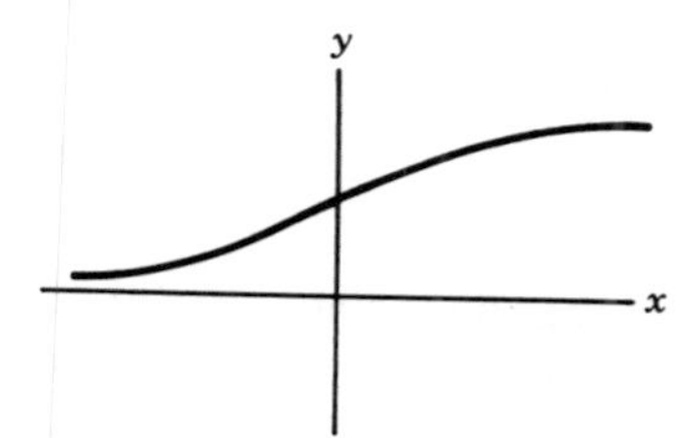

Figure 3.1. y is a continuous nondecreasing function of x.

or

$$F_Y(y) = F_X(x) \quad \text{where} \quad y = y(x). \tag{3.4}$$

The relationship between the probability distribution functions may be determined by differentiating Eq. (3.4):

$$f_Y(y)\frac{dy}{dx} = f_X(x). \tag{3.5}$$

Suppose that $y(x)$ is a nonincreasing function of x (Figure 3.2); then

$$P\{y(X) \le y(x)\} = P\{X \ge x\} = 1 - P\{X < x\} \tag{3.6}$$

since

$$P\{X \ge x\} + P\{X < x\} = 1. \tag{3.7}$$

The cumulative distribution function for y is therefore

$$F_Y(y) = 1 - F_X(x) \tag{3.8}$$

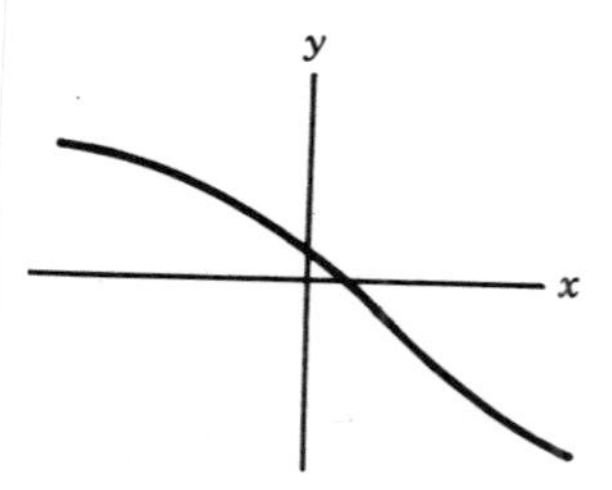

Figure 3.2. y is a nonincreasing function of x.

and

$$f_Y(y)\frac{dy}{dx} = -f_X(x). \tag{3.9}$$

The probabilities in Eq. (3.9) are here nonnegative since dy/dx is negative. The relationship between the pdf's of x and y for both cases can be combined in one equation as

$$f_Y(y)\left|\frac{dy}{dx}\right| = f_X(x). \tag{3.10}$$

Given that x is a random variable with pdf $f_X(x)$ and $y = y(x)$, then

$$f_Y(y) = f_X(x)\left|\frac{dx}{dy}\right| = f_X(x)\left|\frac{dy}{dx}\right|^{-1}. \tag{3.11}$$

Equation (3.11) is also written

$$|f_X(x)\,dx| = |f_Y(y)\,dy|, \tag{3.12}$$

reflecting the fact that all the values of x in dx map into values of y in dy (Figure 3.3). We now give some simple examples.

Suppose that x is a random variable on (0, 1),

$$0 \le x < 1 \quad \text{with} \quad f_X(x) = \frac{4}{\pi}\frac{1}{1+x^2} \tag{3.13}$$

and $y = 1/x$, $1 < y < \infty$; then

$$\frac{dy}{dx} = -\frac{1}{x^2} = -y^2$$

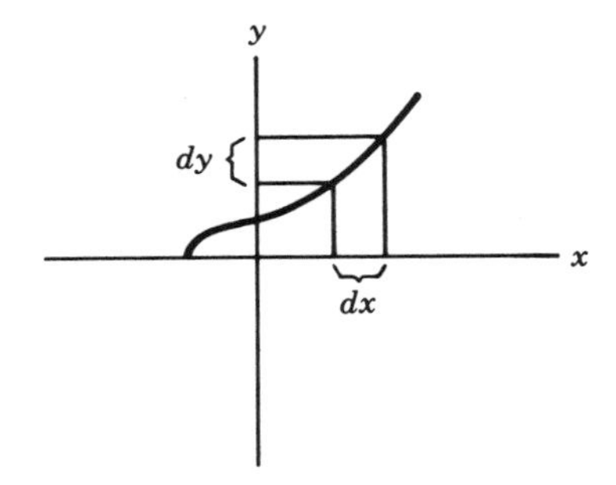

Figure 3.3. Values of x in dx map into values of y in dy.

and

$$f_Y(y) = \left(\frac{4}{\pi}\right)\frac{1}{1+x^2}\,y^{-2} = \left(\frac{4}{\pi}\right)\frac{1}{1+1/y^2}\,\frac{1}{y^2} = \frac{4}{\pi}\frac{1}{1+y^2}. \tag{3.14}$$

The pdf $f_Y(y)$ is a different distribution. In this case, however, it has the same functional form as $f_X(x)$, but on a different range.

As another example, consider the linear transformation $y = a + bx$. Now

$$f_Y(y) = |b^{-1}|f_X\left(\frac{y-a}{b}\right). \tag{3.15}$$

In particular, let $f_X(x) = 1$ for x on $[0, 1]$; then

$$f_Y(y) = \begin{cases} b^{-1}, & a \le y \le a+b \quad \text{for} \quad b > 0 \\ -b^{-1}, & a+b \le y \le a \quad \text{for} \quad b < 0. \end{cases} \tag{3.16}$$

Equation (3.15) can be used to prove a relation asserted in Section 2.8. The characteristic function of $y(x)$ is

$$c_y(t) = \int_{-\infty}^{\infty} e^{iyt} f_Y(y)\,dy = b^{-1}\int_{-\infty}^{\infty} e^{iyt} f_X\left(\frac{y-a}{b}\right)dy. \tag{3.17}$$

By changing variables $u = (y-a)/b$, the integral becomes

$$c_y(t) = \int_{-\infty}^{\infty} e^{i(a+bu)t} f_X(u)\,du = e^{iat}c_x(bt). \tag{3.18}$$

For the case when $a = 0$ and $b = 1/n$, Eq. (3.18) reduces to

$$c_y(t) = c_x\left(\frac{t}{n}\right). \tag{3.19}$$

Suppose x is distributed normally with mean 0 and variance 1:

$$f_X(x) = \phi'(x\,|\,0, 1) = \frac{1}{\sqrt{2\pi}}\exp\left[-\frac{x^2}{2}\right], \qquad -\infty < x < \infty, \tag{3.20}$$

and y is a linear transformation of x, $y = \sigma x + \mu$. Then

$$f_Y(y) = \frac{1}{\sqrt{2\pi}\sigma} \exp\left[-\frac{1}{2}\left(\frac{y-\mu}{\sigma}\right)^2\right]. \tag{3.21}$$

The random variable y is also normally distributed, but its distribution function is centered on μ and has variance σ^2.

In the discussion so far we have talked about transforming a random variable x having any distribution into a random variable y. Because conventional random number generators yield values uniform on (0, 1), the transformation from that case is particularly important. The pdf for a uniform random variable on (0, 1) is

$$f_\xi(\xi) = \begin{cases} 0 & \text{for } \xi < 0 \text{ or } \xi > 1 \\ 1 & \text{otherwise.} \end{cases} \tag{3.22}$$

Now consider as an example the family of functions

$$y = \xi^r. \tag{3.23}$$

Then the probability distribution function for y is

$$f_Y(y) = \left|\frac{1}{r}\right| y^{1/r-1} \begin{cases} 0 < y < 1 & \text{if } r > 0 \\ 1 < y < \infty & \text{if } r < 0. \end{cases} \tag{3.24}$$

If $r > 1$, the power of y will be negative, for example,

$$r = 2 \Rightarrow f_Y(y) = \tfrac{1}{2} y^{-1/2}.$$

This pdf diverges at the origin; distributions can be singular. As $r \to \infty$, the pdf becomes arbitrarily close to $f_Y(y) = y^{-1}$ but never reaches it since $f_Y(y)$ must be integrable. If it is necessary to sample a power law, Eq. (3.24), on $0 < y < 1$ or $y > 1$, then the transformation given above, Eq. (3.23), may be used. The functions in Eq. (3.24) will also be useful in evaluating integrals of singular functions by sampling singular pdf's. What is meant by this will be explained more fully in Chapter 4.

Another useful transformation is

$$y = -\log \xi, \qquad y \text{ on } (0, \infty). \tag{3.25}$$

The transformation can be written $\xi = e^{-y}$ so that the pdf for y is

$$f_Y(y) = f_\xi(\xi)\left|\frac{dy}{d\xi}\right|^{-1},$$

$$\left|\frac{dy}{d\xi}\right| = \frac{1}{\xi} = e^y,$$

so finally

$$f_Y(y) = e^{-y}. \tag{3.26}$$

That is, the random variable which is the natural logarithm of a uniform random variable is distributed exponentially.

Specific Algorithms

We now consider the problem of finding an algorithm to sample a specified function. This is usually the form in which the problem is posed. For univariate distributions, there is a general inversion technique that may be justified as follows.

Let $y(x)$ be an increasing function of x. The cumulative distribution function of y may be determined from (3.4). If ξ is uniform, its cumulative distribution function is

$$F_\xi(\xi) = \begin{cases} 0, & \xi < 0 \\ \xi, & 0 \le \xi \le 1 \\ 1, & \xi \ge 1. \end{cases} \tag{3.27}$$

Therefore on (0, 1) the cumulative distribution function for y is determined by solving the equation

$$F_Y(y) = \xi \tag{3.28}$$

for y.

Suppose the pdf for y is given by

$$f_Y(y) = \lambda e^{-\lambda y}, \qquad 0 < y < \infty, \tag{3.29}$$

then from Eqs. (3.28) and (3.29)

$$F_Y(y) = \int_0^y \lambda e^{-\lambda u}\, du = 1 - e^{-\lambda y} = \xi. \tag{3.30}$$

This yields

$$y = -\frac{1}{\lambda}\log(1-\xi), \tag{3.31}$$

an increasing function of ξ. The expression in Eq. (3.31) is computationally equivalent* to $-(1/\lambda)\log(\xi)$. This is true since, if ξ is uniform on (0, 1), then $1-\xi$ is also uniform on (0, 1). The decision about which form to use depends both on whether $f_Y(y)$ is increasing or decreasing and on convenience.

As another example, let

$$f_Y(y) = \frac{2}{\pi}\frac{1}{1+y^2}, \qquad 0 < y < \infty; \tag{3.32}$$

then the cumulative distribution function is

$$F_Y(y) = \int_0^y \frac{2}{\pi}\frac{1}{1+u^2}\,du = \frac{2}{\pi}\tan^{-1} y = \xi.$$

Solving this equation for y yields

$$y = \tan\frac{\pi}{2}\xi. \tag{3.33}$$

If a random variable y is required having pdf $[\pi(1+y^2)]^{-1}$ on $(-\infty, \infty)$, this may be accomplished by assigning a sign to y randomly:

$$\text{if } \xi_1 < \tfrac{1}{2}, \qquad y = -\tan\frac{\pi}{2}\xi_2,$$

$$\text{if } \xi_1 > \tfrac{1}{2}, \qquad y = +\tan\frac{\pi}{2}\xi_2.$$

It is left to the reader to show that if the pdf of y is

$$f_Y(y) = \frac{1}{\pi}\frac{1}{1+y^2}, \qquad -\infty < y < \infty,$$

*By *computationally equivalent* we do not mean that the value of y is the same in both cases. Rather, the distributions are the same and both give statistically equivalent results when used in a Monte Carlo calculation.

then y may be obtained from the transformation

$$y = \tan\frac{\pi}{2}(2\xi - 1).$$

Another useful pdf is

$$f_R(r) = r\exp[-\tfrac{1}{2}r^2], \qquad 0 < r < \infty, \tag{3.34}$$

whose cumulative distribution function is given by

$$F_R(r) = \int_0^r u\exp[-\tfrac{1}{2}u^2]\,du = 1 - \exp[-\tfrac{1}{2}r^2] = \xi.$$

A random variable that is distributed as in Eq. (3.34) is then

$$r = [-2\log(1-\xi)]^{1/2}. \tag{3.35}$$

It is frequently necessary to sample a gaussian

$$\phi'(y \mid 0, 1) = \frac{1}{\sqrt{2\pi}}\exp[-\tfrac{1}{2}y^2], \qquad -\infty < y < \infty, \tag{3.36}$$

in Monte Carlo calculations. In practice, it is easier to sample two independent gaussian random variables together, that is,

$$f(y_1, y_2) = \phi'(y_1 \mid 0, 1)\phi'(y_2 \mid 0, 1) = \frac{1}{2\pi}\exp[-\tfrac{1}{2}(y_1^2 + y_2^2)].$$

Now, the equation can be transformed into the independent polar coordinates r and ϕ by the transformation

$$y_1 = r\cos\phi,$$
$$y_2 = r\sin\phi,$$

and rewritten

$$\phi'(y_1)\phi'(y_2)\,dy_1\,dy_2 = (\exp[-\tfrac{1}{2}r^2]r\,dr)\left(\frac{1}{2\pi}\,d\phi\right). \tag{3.37}$$

The angle ϕ is distributed uniformly on $(0, 2\pi)$ and may be sampled by

$$\phi = 2\pi\xi_2.$$

The probability distribution function for r is the same as that introduced in Eq. (3.34), so r can be sampled as in (3.35). The two independent gaussian random variables become

$$y_1 = [-2 \log \xi_1]^{1/2} \cos 2\pi\xi_2, \tag{3.38}$$
$$y_2 = [-2 \log \xi_1]^{1/2} \sin 2\pi\xi_2;$$

this is known as the *Box–Muller* method (though it was invented by Wiener[2]). The equations can be linearly transformed to any μ and any σ as described above [see Equation (3.21)]. A good subroutine may be programmed using the Box–Muller method; however, as written the method is convenient but slow. One advantage is that it permits the sampling of a gaussian random variable in one FORTRAN expression.

An approximate gaussian random variable may also be generated by invoking the central limit theorem. By sampling N uniform random variables $\xi_1, \xi_2, \ldots, \xi_N$ and forming the sum

$$y = \sqrt{12/N}\left(\sum_{k=1}^{N} \xi_k - \frac{N}{2}\right), \tag{3.39}$$

a variable with mean 0 and variance 1 is generated. The central limit theorem asserts that this will be nearly gaussian for large N. A value of $N = 12$ appears to be sufficiently large for many purposes and avoids the evaluation of the factor $\sqrt{12/N}$.

3.2. NUMERICAL TRANSFORMATION

In the preceding examples an expression for sampling a random variable was derived by applying

$$F_Y(y) = \xi.$$

This equation can always be used if $y(\xi)$ is an increasing function, but in practice solving it may require a slow iteration procedure. It may be a transcendental equation that must be solved anew for each value of ξ.

An attempt to generate a gaussian random variable by the equation

$$\phi(x \mid 0, 1) = \xi$$

requires solving the expression for the error function

$$\frac{1}{\sqrt{2\pi}} \int_0^x \exp\left[-\frac{u^2}{2}\right] du = \operatorname{erf}(x) = \xi$$

for x. The method of setting a uniform random variable equal to the cumulative distribution function is useful only if the resulting equation is economically solvable. Finally it may be that $f(y)$ is known only numerically (e.g., from experimental data).

A general, straightforward, and economical approximation to the required inversion may be obtained if the values of y for which the cumulative distribution function has specified values can be computed in advance. That is, let

$$F(y_n) = \int_0^{y_n} f_Y(y)\, dy = \frac{n}{N}, \qquad n = 0, 1, 2, \ldots, N, \tag{3.40}$$

where $y_0 = 0$ and $y_N = Y$. As before we must solve

$$F(y) = \xi$$

for y. Find n such that

$$\frac{n}{N} < \xi < \frac{n+1}{N}.$$

The value for $y(\xi)$ may be calculated by linear interpolation

$$y(\xi) = y_n + (y_{n+1} - y_n)u, \tag{3.41}$$

where

$$u = N\xi - n, \qquad 0 < u < 1.$$

This method corresponds to approximating a pdf by a piecewise-constant function with the area of each piece a fixed fraction as in Figure 3.4. It is most accurate when the pdf is large and is least accurate when the pdf is

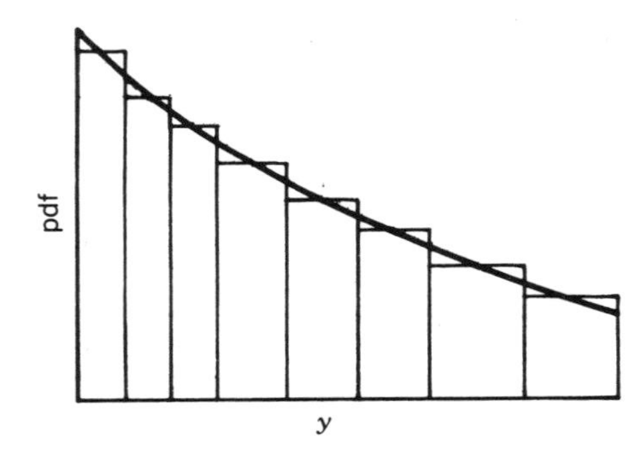

Figure 3.4. Approximating a numerical $f(y)$ by a piecewise-constant function.

small. Rather than linearly interpolating, the value of y may simply be set to the nearest y_n. This approximates the original $f_Y(y)$ as a discrete distribution.

3.3. SAMPLING DISCRETE DISTRIBUTIONS

According to Eq. (3.22) the distribution function for a uniform random variable is

$$f_\xi(\xi) = 1, \qquad 0 \le x \le 1.$$

Using also Eq. (2.18), we have that

$$P\{0 \le x_1 < \xi \le x_2 \le 1\} = F_\xi(x_2) - F_\xi(x_1) = x_2 - x_1. \tag{3.42}$$

The chance that ξ lies in an interval $[x_1, x_2]$ of $(0, 1)$ is equal to the length of the interval. Suppose we have a class of events E_k with probabilities f_k and we wish to sample one at random. We may generate a uniform variable ξ and, if it lies in an interval of length f_k on $(0, 1)$, assign event k to that trial. Better, since $\sum f_k = 1$, it is possible to take the interval $(0, 1)$ and exhaust it by dividing it into segments each of which has a length equal to some f_l (Figure 3.5). The interval into which a ξ falls determines the identity of the event.

A uniform random variable is generated, and the smallest l is found for which the sum of the f_k is greater than the random number; that is,

$$\sum_{k=0}^{l-1} f_k < \xi \le \sum_{k=0}^{l} f_k. \tag{3.43}$$

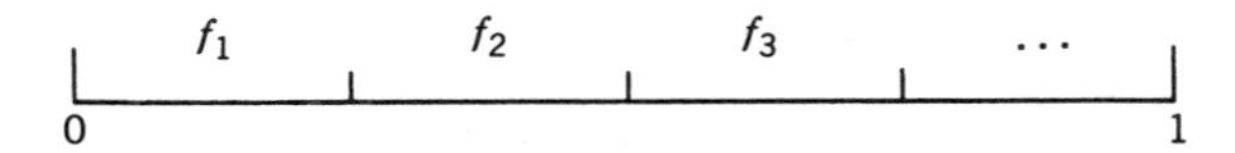

Figure 3.5. Dividing the interval (0, 1) into segments of length f_l.

(When $l=0$, the sum is defined to equal 0.) Whenever $0<\xi<f_1$, event 1 takes place; if $f_1<\xi<f_1+f_2$, event 2 takes place; and so on.

Thus if we must choose between equally likely events, we may consider the first to be selected if some $\xi<\frac{1}{2}$; otherwise, the second is selected. To select three events with probabilities $\frac{1}{2}$, $\frac{1}{4}$, and $\frac{1}{4}$, we choose the first if $\xi<\frac{1}{2}$, the second if $\xi<\frac{3}{4}$, and the third otherwise.

Suppose we must choose among K equally likely events,

$$f_k=\frac{1}{K}, \qquad k=1,2,\ldots,n.$$

The sums in Eq. (3.43) are formed

$$\sum_{k=0}^{l-1} f_k<\xi\leq\sum_{k=0}^{l} f_k$$

which reduce to

$$\frac{l-1}{K}<\xi\leq\frac{l}{K},$$

$$l-1<\xi K\leq l.$$

The appropriate value of l for a particular ξ is then

$$l=[\xi K]+1,$$

where $[u]$ indicates truncation to the largest integer less than u.

In searching for an index l satisfying (3.43), a binary search is strongly recommended when the total number of intervals is large and if $\sum f_k$ does not converge fast. If a serial search is to be used and the index can be arranged at our disposal, then the index with the largest probability should be put in the first place and so on, to reduce the average time of searching.

An illustration of this is found in sampling the geometric distribution

$$f_k = 2^{-k}, \qquad k = 1, 2, \ldots,$$

which may be sampled by the FORTRAN loop

```
      FK = 0.5
      X = RN(D)
      DO 100 K = 1,9999
         IF(X.LE.FK) GO TO 120
100   FK = 0.5*FK
120   CONTINUE
```

Since the probabilities decrease, the chances are good that the loop terminates quickly. In fact two tests are required, on the average, before it does. It is possible to sample k directly [i.e., by solving Eq. (3.43) explicitly] to get (in FORTRAN)

```
K = -ALOG(RN(D))/ALOG(2.) + 1.
```

The decision on which method to use to sample the geometric distribution depends on the computer used. If evaluations of logarithms is slow, it is better to use the multiplicative method shown first.

On the other hand, for sampling

$$f_k = (1 - \alpha)\alpha^{k-1}; \qquad k = 1, 2, \ldots,$$

the expected number of passes through a loop like that of the code given above is $1/(1 - \alpha)$, which is large for α close to 1. A binary search can be carried out in time proportional to $-\log(1 - \alpha)$, but the computation by way of $-\log(\xi)/\log(\alpha)$ takes constant time.

Further illustrations of the sampling of discrete distributions will be given in the discussion of other methods. In particular, the composition method (Section 3.4) involves sampling discrete marginal distributions and continuous (or discrete) conditional distributions.

A mixed distribution has an F that is partly continuous, but with step discontinuities. An example is shown in Figure 3.6. It may be sampled by mapping if care is taken to map ξ to the appropriate part of F.

Consider

$$F_X(x) = \begin{cases} 0 & \text{for } x < 0 \\ 1 - \frac{1}{2}e^{-\lambda x} & \text{for } x > 0. \end{cases} \tag{3.44}$$

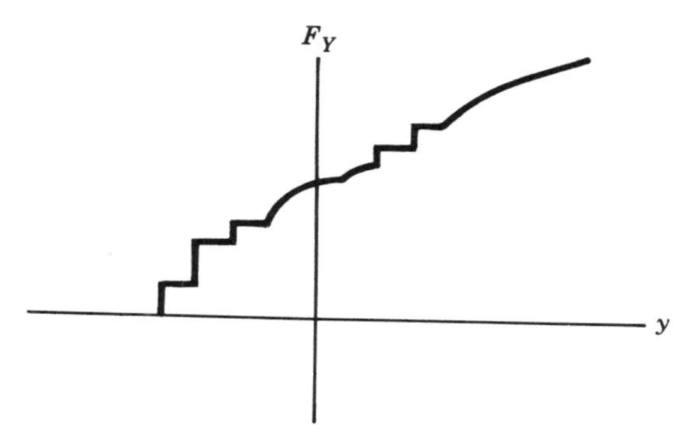

Figure 3.6. A mixed distribution function.

The step of $\frac{1}{2}$ at $x=0$ indicates that the discrete value $x=0$ occurs with probability $\frac{1}{2}$; values of $x>0$ are distributed continuously from zero to infinity. Negative values of x never occur. To sample this, we select ξ and set

$$x=\begin{cases} 0 & \text{if } \xi \leq \frac{1}{2} \\ -\log(2(1-\xi))/\lambda & \text{otherwise.} \end{cases} \tag{3.45}$$

That is, we solve $F_X(x)=\xi$ when $\xi>\frac{1}{2}$.

In summary, suppose that $y(x)$ is a discrete random variable. A value of y is sampled by generating a uniform random variable and deciding what value of y contains ξ in its range:

$$P(\xi \text{ in a particular range}) = \text{length of discrete segment.}$$

This procedure was demonstrated in Eq. (3.43). If F also contains continuous sections, Eq. (3.28) must be solved in these sections.

3.4. COMPOSITION OF RANDOM VARIABLES

We have indicated that transforming or mapping random variables may lead to unpleasantly complicated equations to solve numerically. Another technique for generating random variables having a required distribution is to take two or more different (usually independent) random variables drawn from known distributions and combine them in interesting ways. The Box–Muller method of Eq. (3.38) is in fact an example of sampling by composition. The simplest example is simply to add two independent random variables.

3.4.1. Sampling the Sum of Two Uniform Random Variables

Let x, y be uniform on (0, 1) and $z = x + y$; then

$$F(z) = P\{x + y < z\} \tag{3.46}$$

is the area under the line $z = x + y$ within the unit square as shown in Figure 3.7. Consider the case where $x + y < 1$. Geometrically $F(z)$ is seen (Figure 3.8) to be the area of the triangle with sides equaling z:

$$F(z) = \tfrac{1}{2}z^2.$$

For $x + y > 1$, (refer to Figure 3.9) the cumulative distribution function is

$$F(z) = 1 - \tfrac{1}{2}(2 - z)^2.$$

The corresponding pdf's are

$$f(z) = \begin{cases} z, & 0 < z < 1 \\ 2 - z, & 1 \le z < 2, \end{cases} \tag{3.47}$$

which when taken together gives the pdf in Figure 3.10. It is left as an exercise for the reader to find the pdf when two uniform random variables are multiplied.

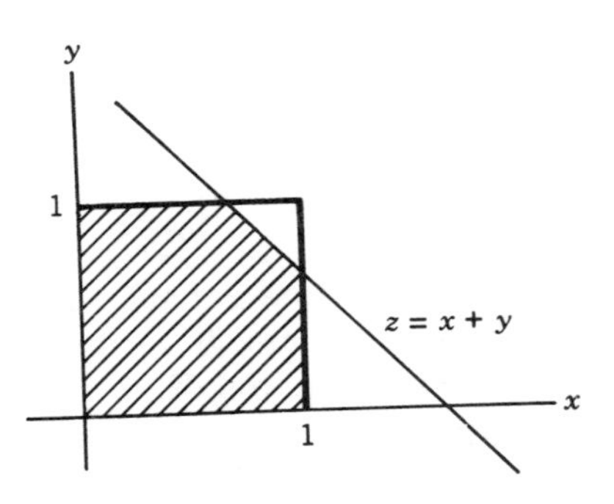

Figure 3.7. The sum of two random variables.

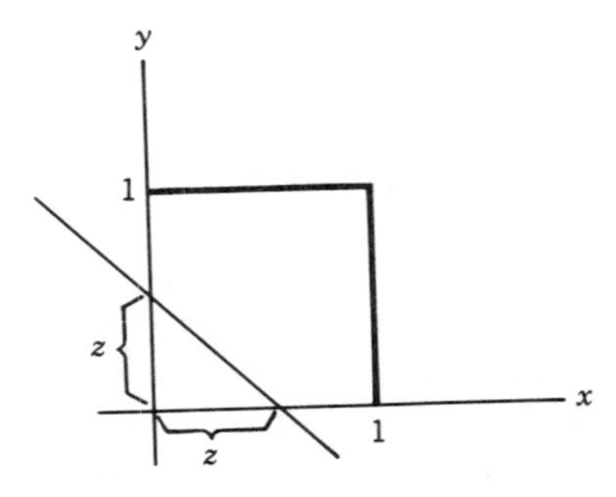

Figure 3.8. $F(z)$ when $x + y < 1$.

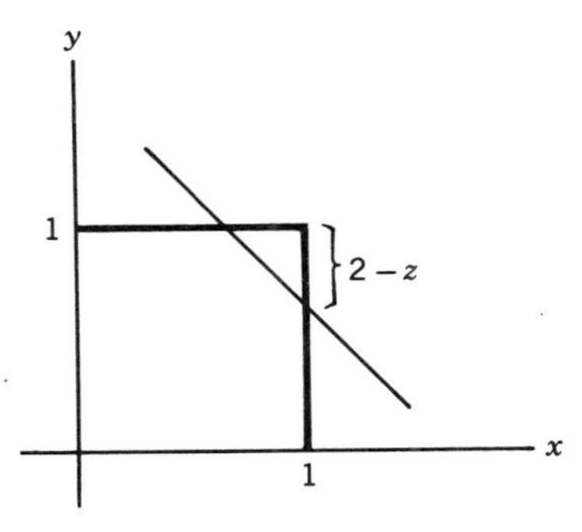

Figure 3.9. $F(z)$ when $x + y > 1$.

3.4.2. Sampling a Random Variable Raised to a Power

Another form of composition is illustrated as follows. Let $x_1, x_2, \ldots, x_n$ be drawn independently from the cumulative distribution functions $F_1(x_1), F_2(x_2), \ldots, F_n(x_n)$. Set z to be the largest of the x_i,

$$z = \max\{x_1, x_2, \ldots, x_n\}. \tag{3.48}$$

What is the distribution function? The following statement holds:

$$Z = \max\{x_1, x_2, \ldots, x_n\} \le z \tag{3.49}$$

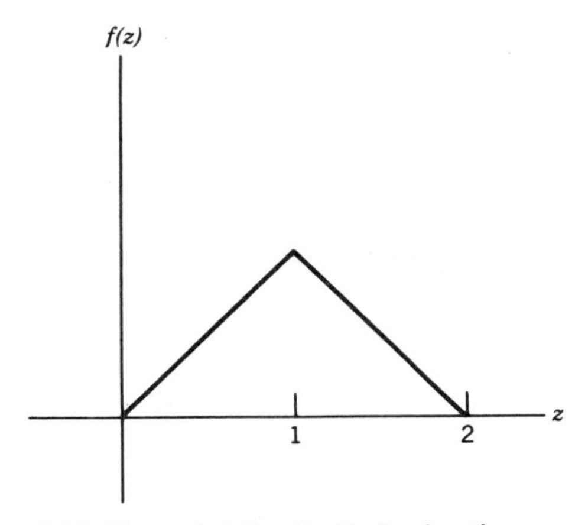

Figure 3.10. The probability distribution function $z = x + y$.

if and only if

$$x_1 \le z \quad \text{and} \quad x_2 \le z \quad \text{and} \quad \cdots \quad \text{and} \quad x_n \le z.$$

Since the x_i are all independently distributed,

$$P\{Z \le z\} = P\{x_1 \le z\} P\{x_2 \le z\} \cdots P\{x_n \le z\}$$

or

$$P\{Z \le z\} = \prod_{i=1}^{n} F_i(z). \tag{3.50}$$

A similar proof holds for the smallest of n random variables.

Suppose z equals the maximum of k uniform random variables

$$z = \max\{\xi_1, \xi_2, \ldots, \xi_k\}; \tag{3.51}$$

then

$$F(z) = \prod_{i=1}^{k} F_i(z) = z^k \tag{3.52}$$

since

$$F_i(z) = z$$

for a uniform random variable. The pdf corresponding to $F(z)$ is

$$f(z) = kz^{k-1}, \qquad 0 < z < 1. \tag{3.53}$$

A FORTRAN routine for carrying out (3.51) is

```
      PMAX = 0.
      DO 50 IZ = 1,K
      PMAX = AMAX1(RN(D),PMAX)
   50 CONTINUE
```

It is also possible to sample a z by transformation:

$$\begin{aligned} F(z) &= z^k = \xi, \\ z &= \xi^{1/k}, \end{aligned} \tag{3.54}$$

which in a FORTRAN statement is written

```
z = RN(D)**(1./K).
```

In a computer calculation, which method will be more efficient? Though the sampling method in Eq. (3.54) may be written in one FORTRAN line, the statement invokes a lengthy process requiring a logarithm and exponential for the actual computation. Therefore, method (3.51) is faster for small values of k. As the value of k increases, more and more random numbers must be computed in the loop while the computation time of Eq. (3.54) will be independent of the value of k. For larger values of k, it is preferable to employ the second method. The choice of algorithm for sampling a random variable should be guided by what is most efficient. The decision is machine and compiler dependent.

3.4.3. Sampling the Distribution $f(z) = z(1 - z)$

Let z equal the middle value of three random numbers,

$$z = \mathrm{mid}(\xi_1, \xi_2, \xi_3); \tag{3.55}$$

the probability distribution function for z is then

$$f(z) = 6z(1 - z).$$

This result may be easily derived. Assume that $\xi_1 < \xi_2 < \xi_3$, so that $z = \xi_2$. The probability that ξ_2 is in the range $d\xi_2$ and that $\xi_1 < \xi_2$ and $\xi_3 > \xi_2$ is

$$f(\xi_2)\, d\xi_2 = d\xi_2\, P(\xi_1 < \xi_2) P(\xi_3 > \xi_2) = d\xi_2\, \xi_2(1 - \xi_2).$$

The probability that the middle value is found in any possible range $d\xi$ becomes

$$\int_0^1 f(\xi)\, d\xi = \int_0^1 \xi(1 - \xi)\, d\xi = \frac{1}{6}.$$

Taking account of the six possibilities for the arrangement of ξ_1, ξ_2, ξ_3, we get for the total

$$f(z) = 6z(1 - z). \tag{3.56}$$

The corresponding cumulative distribution function is

$$P(\text{middle } Z < z) = F(z) = 3z^2 - 2z^3.$$

3.4.4. Sampling the Sum of Several Arbitrary Distributions

Often a distribution function that must be sampled has (or may be written in) the form

$$\begin{gathered} f_X(x) = \sum_n \alpha_n g_n(x) \\ \alpha_n \geq 0, \qquad g_n(x) \geq 0, \\ \int g_n(x) \neq 1, \end{gathered} \tag{3.57}$$

but as usual

$$\int f_X(x)\, dx = 1.$$

We now demonstrate how such a sum of terms may be sampled by random selection of a single term of the sum followed by sampling the distribution of that term. Consider a set of functions $h_n(x)$ and coefficients β_n satisfying

$$\begin{gathered} h_n(x) \geq 0, \qquad \int h_n(x)\, dx = 1, \\ \beta_n \geq 0, \qquad \sum \beta_n = 1. \end{gathered} \tag{3.58}$$

The β_n are effectively probabilities for the choice of an event n. Let us select event m with probability β_m. Then sample X from $h_m(x)$ for that m. What is the distribution of the values of x that result? That is, what is the probability that $X \leq x$? The probability that m is chosen *and* $X \leq x$ is $\beta_m \int_0^x h_m(t)\, dt$. Since different m are mutually exclusive, the total probability that $X \leq x$ is the sum over all m is

$$P\{X \leq x\} = \sum_m \beta_m \int_0^x h_m(t)\, dt. \tag{3.59}$$

The probability density function that results is just

$$h(x) = \sum_m \beta_m h_m(x). \tag{3.60}$$

Again we emphasize that although a *single* h_m is sampled in one trial, the result, taking into account that *any* m can be used, is to sample the entire sum.

Returning to Eq. (3.57) we observe that it may be rewritten

$$f_X(x) = \sum_n \alpha_n \left[\int g_n(u)\,du\right]\left[\frac{g_n(x)}{\int g_n(w)\,dw}\right]. \tag{3.61}$$

We identify

$$\beta_n = \alpha_n \int g_n(u)\,du \tag{3.62}$$

and

$$h_n(x) = \frac{g_n(x)}{\int g_n(w)\,dw}$$

and note that the conditions (3.58) are satisfied.

An example will illustrate the ideas. Consider the pdf

$$f_X(x) = \tfrac{3}{5}(1 + x + \tfrac{1}{2}x^2), \qquad 0 < x < 1; \tag{3.63}$$

the sum is rewritten

$$f_X(x) = \tfrac{3}{5}\cdot 1 + \tfrac{3}{5}\cdot\tfrac{1}{2}\cdot 2x + \tfrac{3}{5}\cdot\tfrac{1}{2}\cdot\tfrac{1}{3}\cdot 3x^2$$

so that

$$\begin{aligned} \beta_1 &= \tfrac{3}{5}, & h_1 &= 1,\\ \beta_2 &= \tfrac{3}{10}, & h_2 &= 2x,\\ \beta_3 &= \tfrac{1}{10}, & h_3 &= 3x^2, \end{aligned}$$

and

$$\sum \beta_n = 1.$$

The value of n is chosen with probability β_n, that is,

0. sample ξ_0;
1. if $\xi_0 \leq \frac{6}{10}$, $n = 1$;
2. else if $\xi_0 \leq \frac{9}{10}$, $n = 2$;
3. else $n = 3$.

Once a value for n is selected, the appropriate h_n is sampled for x:

if $n = 1$, set $x = \xi_1$;
if $n = 2$, set $x = \max(\xi_1, \xi_2)$;
if $n = 3$, set $x = \max(\xi_1, \xi_2, \xi_3)$

since the distribution functions for x are all power laws.

Another example is the pdf

$$f_X(x) = \frac{1}{4}\left[\frac{1}{x^{1/2}} + \frac{1}{(1-x)^{1/2}}\right], \qquad 0 < x < 1. \tag{3.64}$$

This may also be sampled using the algorithm suggested above by rewriting it in the form

$$f_X(x) = \frac{1}{2}\frac{1}{2x^{1/2}} + \frac{1}{2}\frac{1}{2(1-x)^{1/2}}$$

so that

$$\beta_1 = \frac{1}{2}, \qquad h_1(x) = \frac{1}{2x^{1/2}},$$

$$\beta_2 = \frac{1}{2}, \qquad h_2(x) = \frac{1}{2(1-x)^{1/2}}.$$

The cumulative distribution function corresponding to $h_1(x)$ is

$$H_1(x) = x^{1/2} = \xi$$

or

$$x = \xi^2;$$

similarly the cumulative distribution function corresponding to $h_2(x)$ is

$$H_2(x) = 1 - (1 - x^2)^{1/2}$$

and

$$x = 1 - (1 - \xi)^2.$$

But this equation may be simplified by remembering the argument that $1 - \xi$ is uniform on (0, 1) if ξ is, so

$$x = 1 - \xi^2.$$

The sampling algorithm becomes

1. $x = \xi_1^2$;
2. if $\xi_2 > \frac{1}{2}$, $x = 1 - x$ (since $\beta_n = \frac{1}{2}$). (3.65)

This is a simple method for sampling a complicated pdf.

3.5. REJECTION TECHNIQUES

There is a kind of composition method that leads to very general techniques for sampling any probability distribution. It has one new feature, namely, that a trial value for a random variable is selected and proposed. This value is subjected to one or more tests (involving one or more other random variables) and it may be accepted, that is, used as needed, or rejected. If it is rejected, the cycle of choosing and testing a trial value is repeated until an acceptance takes place. An important property of the method is that the normalization of the density need not be known explicitly to carry out the sampling.

A disadvantage is that it may have low *efficiency*, that is, many values are rejected before one is accepted. It is an interesting technical challenge to devise efficient rejection techniques for varieties of distributions.

The idea is best conveyed by a simple general case and a specific example.

Suppose we want to sample a complicated pdf on (0, 1), and that the "trial value" x_0 is chosen uniformly between 0 and 1. The test to which x_0 is subjected must accept few values when $f(x_0)$ is small (near $x_0 = 1$ in Figure 3.11) and most values when $f(x_0)$ is large (near $x_0 = 0$). This is accomplished by accepting x_0 with probability proportional to $f(x_0)$. In

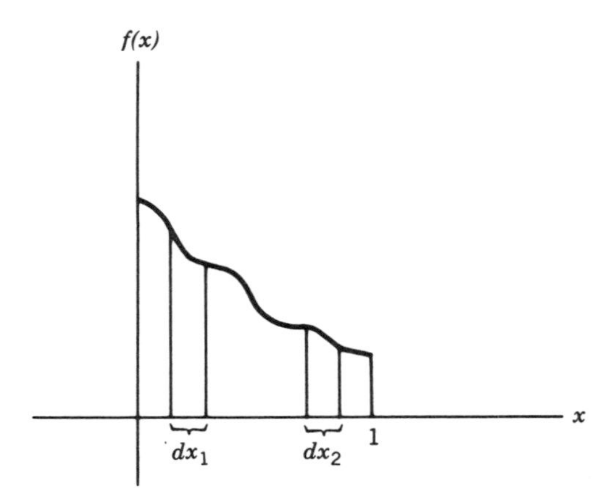

Figure 3.11. A complicated pdf that can be sampled by rejection techniques.

Figure 3.11, trial values occur equally often in dx_1 and dx_2 $(=dx_1)$, but we accept a larger fraction of those in dx_1 where $f_X(x)$ is larger.

The pdf in the figure has the property that

$$f_X(x) \le f(0). \tag{3.66}$$

A test that meets exactly our requirement that the density of accepted x_0 is $f(x_0)$ is

$$\text{Accept } x_0 \quad \text{if} \quad \xi_2 \le \frac{f(x_0)}{f(0)}. \tag{3.67}$$

Put another (geometric) way, points are chosen uniformly in the smallest rectangle that encloses the curve $f_X(x)$. The ordinate of such a point is $x_0 = \xi_1$; the abscissa is $x_1 = f(0)\xi_2$. Points lying above the curve are rejected. Points below are accepted; their ordinates x_0 have distribution $f_X(x)$.

The method can be clarified by a concrete example: let

$$f_X(x) = \frac{4}{\pi}\frac{1}{1+x^2}, \qquad 0 < x < 1, \tag{3.68}$$

which is a monotonically decreasing function on (0, 1) (Figure 3.12). A series of random points are generated uniformly on a rectangle enclosing $f_X(x)$. Only those points that lie below the curve $f_X(x)$ are accepted; the

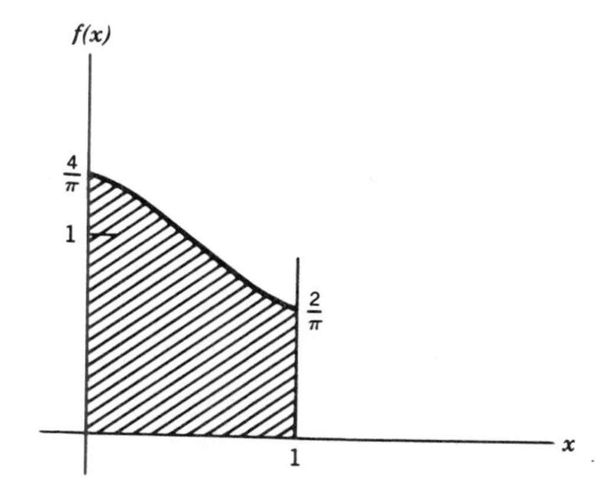

Figure 3.12. The pdf $(4/\pi)[1/(1+x^2)]$.

ordinate is x. The algorithm for accomplishing this is

1. $x_0 = \xi_1$;
2. $x_1 = (4/\pi) \cdot \xi_2$;
3. if $x_1 \le (4/\pi)[1/(1+x_0^2)]$ accept x_0; otherwise repeat from step 1.

The algorithm may be rephrased more succinctly as

1. $x_0 = \xi_1$;
2. if $\xi_2 > 1/(1+x_0^2)$ then repeat from 1

or more efficiently as

1. if $\xi_2(1+\xi_1^2) > 1$ repeat; otherwise $x_0 = \xi_1$.

Let us derive the density function generated by a somewhat wider class of rejection methods. Let a random variable Z have a pdf $g(z)$. This Z is accepted if $\xi_2 \le h(Z) < 1$; else sample another Z. There are two discrete events: a success (Z is accepted) and no success (Z is rejected). The joint probability that $Z < z$ and that $\xi_2 \le h(Z)$ is

$$P\{Z < z \text{ and } \xi_2 \le h(Z)\} = \int_{-\infty}^{z} h(t)g(t)\,dt, \tag{3.69}$$

where $h(t)$ is the probability of success given t and $g(t)\,dt$ is the probability that t is in dt. We may write the joint probability as the product of a marginal probability for success and a conditional probability that $Z < z$:

$$P\{Z < z \text{ and success}\} = P\{\text{success}\}P\{Z < z \mid \text{success}\}.$$

But, $P\{Z < \infty\} = 1$ so that

$$P\{\text{success}\} = \int_{-\infty}^{\infty} h(z)g(z)\, dz.$$

The rejection technique yields a Z only when a success occurs, that is,

$$\begin{aligned} P\{Z < z \mid \text{success}\} &= \text{distribution of } Z\text{'s coming from a rejection algorithm} \\ &= \frac{\int_{-\infty}^{z} h(t)g(t)\, dt}{\int_{-\infty}^{\infty} h(z)g(z)\, dz} \end{aligned} \tag{3.70}$$

therefore, the probability distribution of Z's that results from the rejection technique is

$$f(z) = \frac{h(z)g(z)}{\int_{-\infty}^{\infty} h(t)g(t)\, dt}. \tag{3.71}$$

In a previous example [Eq. (3.68)], the initial value of z is chosen uniformly on (0, 1) so $g(z) = 1$ and

$$h(z) = \frac{1}{1 + z^2}.$$

The rejection technique automatically guarantees that the Z's are sampled from a normalized pdf. As it pertains to Monte Carlo, normalization guarantees that an algorithm delivers *some* value of Z.

The a priori probability of a success is called the efficiency, ϵ, and is given by

$$P\{\text{success}\} = \epsilon = \int_{-\infty}^{\infty} h(z)g(z)\, dz. \tag{3.72}$$

The expected number of trials up to and including a success per accepted random variable is

$$\sum_{k=0}^{\infty}(k+1)(1-\epsilon)^k\epsilon=\frac{1}{\epsilon}. \tag{3.73}$$

A better measure of the usefulness of a rejection technique is the expected number of trials multiplied by the computer time per trial.

When $f(z)$ is given, we set the function $h(z)$ to be

$$h(z)=\frac{f(z)/g(z)}{B_h}, \tag{3.74}$$

where B_h is an upper bound for $f(z)/g(z)$. Therefore, $h(z)\leq 1$. The efficiency for this choice of $h(z)$ becomes

$$\epsilon=\frac{1}{B_h}\int_{-\infty}^{\infty}\left[\frac{f(z)}{g(z)}\right]g(z)\,dz=\frac{1}{B_h} \tag{3.75}$$

since $\int_{-\infty}^{\infty} f(z)\,dz=1$. B_h should be the least upper bound for $f(z)/g(z)$ to maximize the efficiency.

Let us consider again the example in Eq. (3.68)

$$f(z)=\frac{4}{\pi}\frac{1}{1+z^2},$$

where $B_h=4/\pi$. The efficiency of the rejection technique is $\epsilon=\pi/4$ and the expected number of trials up to and including the first success is $4/\pi$. The proposed algorithm for sampling z was

1. $z=\xi_1$;
2. if $\xi_2>1/(1+z^2)$, then repeat from 1; otherwise accept z.

If this algorithm were written efficiently, the computer time per accepted z could be as little as 8 μs on a CDC 6600. An alternative method for sampling $f(z)$ discussed in an earlier lecture is

$$z=\tan\frac{\pi}{4}\xi,$$

which will take approximately 40 μs per z on a CDC 6600. This is a circumstance in which the rejection technique is more efficient than a direct sampling method.

Consider the singular probability distribution function

$$f(z)=\frac{2}{\pi(1-z^2)^{1/2}}, \qquad 0<z<1; \tag{3.76}$$

this function is unbounded and the straightforward rejection technique introduced above is not appropriate here. Instead, write

$$f(z)=\frac{2}{\pi}\frac{1}{(1+z)^{1/2}(1-z)^{1/2}}$$

and let

$$g(z)=\frac{1}{2(1-z)^{1/2}}; \tag{3.77}$$

then $f(z)/g(z)$ becomes

$$\begin{aligned}\frac{f(z)}{g(z)}&=\frac{2}{\pi}\frac{1}{(1+z)^{1/2}(1-z)^{1/2}}\times 2(1-z)^{1/2}\\&=\frac{4}{\pi}\frac{1}{(1+z)^{1/2}}\leq\frac{4}{\pi}.\end{aligned} \tag{3.78}$$

An appropriate choice for $h(z)$ is therefore

$$h(z)=\frac{f(z)/g(z)}{B_h}=\frac{1}{(1+z)^{1/2}}.$$

An algorithm for selecting a z is

1. sample z from $1/2(1-z)^{1/2}$ (i.e., $z=1-\xi_1^2$);
2. if $\xi_2\leq 1/(1+z)^{1/2}$, accept z; else reject and repeat from 1.

Step 2 may be written more efficiently as $\xi_2^2(1+z)\leq 1$ to avoid a square root and a division.

Rejection techniques have been used extensively in Monte Carlo calculations. A particularly famous algorithm is due to von Neumann[3] for sampling the sine and cosine of an angle uniform on $(0, 2\pi)$. If we are interested in the rotation of objects in three-dimensional space, the angle of rotation may be sampled uniformly, but the sine and cosine of the angle must be calculated for the rotation matrix. The straightforward method would set

$$\phi = 2\pi\xi$$

and then calculate $\cos\phi$ and $\sin\phi$. It would be more efficient to have another method of sampling ϕ that gives the sine and cosine directly. Consider the unit quarter circle inscribed inside the unit square as in Figure 3.13. The direction inside the quarter circle is uniformly distributed

$$f(\phi)\,d\phi = \frac{2}{\pi}\,d\phi.$$

The probability of finding an angle in $d\phi$ is proportional to $d\phi$. Values of x and y are chosen uniformly on (0, 1):

1. $x = \xi_1$ and $y = \xi_2$;
2. if $(\xi_1^2 + \xi_2^2 > 1)$, reject and repeat from 1;
3. $\phi = \tan^{-1} y/x$ when $x^2 + y^2 \leq 1$.

The cosine and the sine in terms of x and y are

$$\cos\phi = \frac{x}{(x^2+y^2)^{1/2}} = \frac{\xi_1}{(\xi_1^2+\xi_2^2)^{1/2}}, \tag{3.79a}$$

$$\sin\phi = \frac{y}{(x^2+y^2)^{1/2}} = \frac{\xi_2}{(\xi_1^2+\xi_2^2)^{1/2}}. \tag{3.79b}$$

If ϕ is uniformly distributed on $(0, \pi/2)$, then 2ϕ is uniformly distributed

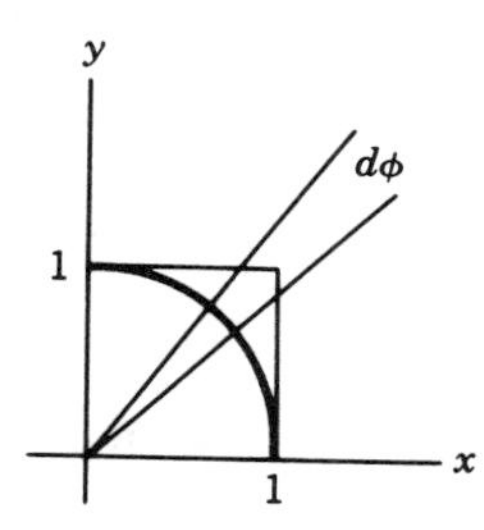

Figure 3.13. Sampling for $\sin\phi$ and $\cos\phi$.

on $(0, \pi)$ and a square root may be avoided by calculating

$$\cos 2\phi = \cos^2 \phi - \sin^2 \phi = \frac{\xi_1^2 - \xi_2^2}{\xi_1^2 + \xi_2^2}, \tag{3.80a}$$

$$\sin 2\phi = 2 \sin \phi \cos \phi = \frac{2\xi_1 \xi_2}{\xi_1^2 + \xi_2^2}. \tag{3.80b}$$

In many calculations (e.g., in selecting direction in three dimensions), however, the $\sin \phi$ and $\cos \phi$ will be used in association with a square root. In these circumstances, there is no disadvantage to using Eq. (3.79) since the square roots can be combined.

If the angle ϕ is to be defined on $(0, 2\pi)$, then Eq. (3.80b) becomes

$$\sin 2\phi = \pm \frac{2\xi_1 \xi_2}{\xi_1^2 + \xi_2^2}, \tag{3.80c}$$

where a third random number is used to choose the sign. An alternative method is to select ϕ uniformly over the circle whose diameter is 1 as shown in Figure 3.14, and set $x = \xi_1 - \frac{1}{2}$ and $y = \xi_2 - \frac{1}{2}$. The values of x and y are accepted only if $x^2 + y^2 < \frac{1}{4}$; and Eqs. (3.79) and (3.80) follow as before.

Another rejection technique, introduced by Kahn,[1] is used to sample a

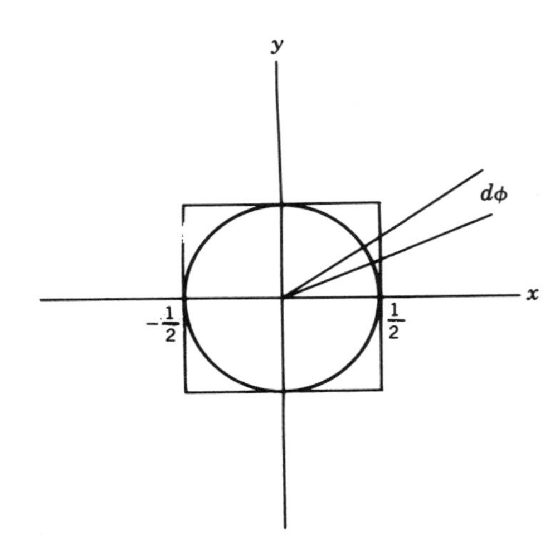

Figure 3.14. Sampling ϕ uniformly over the unit circle.

gaussian. The function

$$\phi'(x) = \frac{1}{\sqrt{2\pi}} \exp\left[-\frac{1}{2} x^2\right], \qquad 0 < x < \infty$$

cannot be bounded within a rectangle since its argument has infinite range. To sample $\phi'(x)$, we must sample from a function defined on the same range. Suppose x has the pdf $\alpha e^{-\alpha x}$, $0 < x < \infty$, and y has the pdf e^{-y}, $0 < y < \infty$; random variables X and Y are chosen by

$$X = -\frac{1}{\alpha} \log \xi_1,$$

$$Y = -\log \xi_2.$$

The value for X is kept if

$$Y > \tfrac{1}{2}(X - \alpha)^2;$$

otherwise new values for X and Y are selected. The probability that X is accepted (success) is

$$\epsilon = \int_{(1/2)(x-\alpha)^2}^{\infty} e^{-y}\, dy \tag{3.81}$$

and the joint probability that $X < x$ and success becomes

$$\begin{aligned} P\{X < x \text{ and success}\} &= \int_0^x \alpha e^{-\alpha t} \int_{(1/2)(t-\alpha)^2}^{\infty} e^{-u}\, du\, dt \\ &= \int_0^x \alpha e^{-\alpha t} e^{-(1/2)(t-\alpha)^2}\, dt \\ &= \alpha e^{-(1/2)\alpha^2} \int_0^x e^{-(1/2)t^2}\, dt. \end{aligned} \tag{3.82}$$

Therefore the probability distribution function for an accepted X is proportional to a gaussian:

$$f(x) = e^{-(1/2)x^2}.$$

A sign to be associated with X could be chosen randomly so that the range is $(-\infty, \infty)$. The proposed algorithm is correct for any positive α, but the efficiency will vary with α:

$$\epsilon = \alpha e^{-(1/2)\alpha^2} \int_0^\infty e^{-(1/2)x^2}\, dx$$

$$= \sqrt{\frac{\pi}{2}}\, \alpha e^{-(1/2)\alpha^2}$$

The maximum efficiency occurs when $\alpha = 1$. The Kahn rejection technique compares favorably with the Box–Muller method [Eq. (3.38)] for sampling a gaussian.

Marsaglia[4] has developed composition methods that are particularly fast and efficient using random access to store fairly large lookup tables. A pdf is written

$$f(x) = \sum b_i g_i(x).$$

The $g_i(x)$ that is sampled the most frequently is chosen to involve a fast table lookup. In the case of the gaussian distribution, $f(x)$ is

$$f(x) = 0.9578\, g_1(x) + 0.0395\, g_2(x) + 0.0027\, g_3(x).$$

$g_1(x)$ and $g_2(x)$ are discrete pdf's that are stored in a table. $g_2(x)$ also involves a rejection. $g_1(x)$ is sampled rapidly and provides x 95% of the time. $g_3(x)$ is used to sample the tail of the gaussian and involves a technique similar to that proposed by Box–Muller. Though the rejection method is much slower than the table lookup, it will occur only 0.27% of the time. Thus overall, Marsaglia's method will generate gaussian variables much more efficiently than a straight rejection technique.

Brown et al.[5] developed a method to sample discrete distributions that is an extension of the Marsaglia algorithms for a vector computer.

The discrete cumulative distribution function is rewritten

$$F(x) = \sum_{k=1}^{N} G(x \mid k) h(k),$$

where $h(k)$ is the discrete (marginal) pdf for deciding k, and $G(x \mid k)$ is the conditional cumulative distribution function for x given k. More specifically,

$$h(k) = \frac{1}{N} \quad \text{for} \quad k = 1, \ldots, N$$

and

$$G(x \mid k) = \begin{cases} 0, & x < x_{1,k} \\ q_k, & x_{1,k} \leq x < x_{2,k}. \\ 1, & x_{2,k} \leq x \end{cases}$$

When k is chosen, $x = x_{1,k}$ with probability q_k and $x = x_{2,k}$ with probability $1 - q_k$. The set $\{q_k, x_{1,k}, x_{2,k}\}$ and number N are calculated from the data $\{f(x_i), x_i\}$, $i = 1, \ldots, N$.

This method is in a sense analogous to the mapping technique of Eq. (3.28) and to its approximation by Eqs. (3.40) and (3.41).

3.6. MULTIVARIATE DISTRIBUTIONS

Except for the bivariate normal distribution, which was introduced as a technical device, our treatment of sampling has been devoted to random variables in one dimension. Multivariate distributions are also important since Monte Carlo is at its best in treating many-dimensional problems. One may use the ideas of marginal and conditional distributions [cf. Eq. (2.1) of Chapter 2] to reduce multivariate to univariate sampling. As usual the point is most easily discussed by an illustration, a pdf of a random variable in $\mathbf{R}^3$:

$$f(\mathbf{x}) = \frac{1}{8\pi} e^{-r}, \tag{3.83}$$

where $r^2 = x^2 + y^2 + z^2$. The volume element dV that is associated with $f(\mathbf{x})$ is $dx\, dy\, dz$. In polar coordinates $dV = r^2\, dr \sin\theta\, d\theta\, d\phi$ and

$$f(\mathbf{x})\, dx\, dy\, dz = \frac{1}{8\pi} e^{-r} r^2\, dr \sin\theta\, d\theta\, d\phi, \tag{3.84}$$

where the polar coordinates are defined as

$$\cos\theta = z/r, \qquad \tan\phi = y/x.$$

Equation (3.84) can be written as the product of three pdf's,

$$f(\mathbf{x})\, dV = \tfrac{1}{2} r^2 e^{-r}\, dr \frac{\sin\theta\, d\theta}{2} \frac{d\phi}{2\pi}, \tag{3.85}$$

and r, θ, and ϕ may be sampled independently. A random variable that will be distributed as $r^2 e^{-r}$ can be sampled* by

$$r = -\sum_{i=1}^{3} \log \xi_i$$
$$= -\log(\xi_1 \cdot \xi_2 \cdot \xi_3).$$

The variable ϕ is distributed uniformly on $(0, 2\pi)$, so

$$\phi = 2\pi\xi_5.$$

The probability distribution for θ can be rewritten

$$\frac{\sin\theta\, d\theta}{2} = -\frac{d(\cos\theta)}{2}.$$

Since $\cos\theta$ is uniformly distributed on $(-1, 1)$, it can be sampled by

$$\cos\theta = 2\xi_4 - 1. \tag{3.86}$$

An alternative derivation of Eq. (3.86) is arrived at by setting $F(\theta)$ equal to a uniform random number,

$$\frac{1}{2}\int_0^{\theta} \sin\theta\, d\theta = \frac{\cos\theta}{2} + \frac{1}{2} = \xi.$$

If ξ is replaced by $1 - \xi_4$, then the last equation reduces to Eq. (3.86). In many applications, it will be x, y, and z that are wanted,

$$z = r\cos\theta,$$
$$x = r\sin\theta\cos\phi,$$
$$y = r\sin\theta\sin\phi.$$

In this case it will be more efficient to calculate $\cos\phi$ and $\sin\phi$ directly, as described in Section 3.5. Also, since $\sin\theta = (1 - \cos^2\theta)^{1/2}$, there will be no loss in efficiency by using the von Neumann rejection technique, which contains square roots [Eq. (3.79)], as was discussed there.

*Note that generally the sum of n independent random variables each with pdf e^{-x} is distributed as $x^{n-1}e^{-x}/(n-1)!$

Singular probability distribution functions are as easily sampled in many dimensions as in one dimension. Let $f(\mathbf{x})\,dV$ be the singular pdf

$$\begin{aligned} f(\mathbf{x})\,dV &= \frac{1}{4\pi}\frac{1}{r^2}\,e^{-r}\,dV \\ &= \frac{1}{4\pi}\frac{1}{r^2}\,e^{-r}r^2\,dr\,d(\cos\theta)\,d\phi \\ &= e^{-r}\,dr\,\frac{d(\cos\theta)}{2}\,\frac{d\phi}{2\pi}. \end{aligned}$$

The variables $\cos\theta$ and ϕ are sampled as described above and r can be sampled by

$$r = -\log \xi.$$

3.7. THE M(RT)² ALGORITHM

The last sampling method we shall discuss is an advanced sampling technique first described in a paper by Metropolis, Rosenbluth, Rosenbluth, Teller, and Teller[6] [M(RT)²].* The method is related to rejection techniques since it involves explicitly proposing a tentative value which may be rejected and because the normalization of the sampled function is irrelevant, we need never know it.

The M(RT)² algorithm is of very great simplicity and power; it can be used to sample essentially any density function regardless of analytic complexity in any number of dimensions. Complementary disadvantages are that sampling is correct only asymptotically and that successive variables produced are correlated, often very strongly. This means that the evaluation of integrals normally produces positive correlations in the values of the integrand with consequent increase in variance for a fixed number of steps as compared with independent samples. Also the method is not well suited to sampling distributions with parameters that change frequently.

The usual description of the M(RT)² method can be found in the papers by Valleau and Whittington,[7] Valleau and Torrie,[8] and Wood and Erpenbeck.[9] Here we shall develop a somewhat more general description.

The method was motivated by an analogy with the behavior of systems

*In the literature, it is often referred to as the Metropolis algorithm.

in statistical mechanics that approach an equilibrium whose statistical properties are independent of the kinetics of the system.

By *system* we mean here simply a point x in a space Ω (typically in $\mathbf{R}^M$) that may be thought of as a possible description of a physical problem. By *kinetics* we mean a stochastic transition that governs the evolution of the system: a probability density function $K(X \mid Y)$ that the evolution of a system known to be at Y will next bring it near X. K may be thought of as a model of the physical process by which a system changes or as a mathematical abstraction. In a Monte Carlo calculation it will play the role of a sampling distribution.

As we shall discuss in detail, one condition that a system evolve toward equilibrium and stay there is, quite simply, that the system be on the average as likely to move into a specific neighborhood of X from a neighborhood of Y as to move exactly in the reverse direction. If the probability density for observing the system near X in equilibrium is $f(X)$, then the kinetics must satisfy

$$K(X \mid Y) f(Y) = K(Y \mid X) f(X). \tag{3.87}$$

This relation is called *detailed balance*.[10] $K(X \mid Y) f(Y)$ is the probability of moving from Y to X expressed as the a priori chance of finding the system near Y (i.e., $f(Y)$) times the conditional probability $[K(X \mid Y)]$ that it will move to X from Y.

In treating a physical system, one usually assumes that $K(X \mid Y)$ is known, and one has the task of finding $f(X)$. The M(RT)^2 algorithm (as in much of Monte Carlo) reverses this: one has the task of finding a convenient and correct kinetics that will equilibrate the system so that the given $f(X)$ turns out to be the chance of observing the system near X.

This turns out to be extremely easy given the elegant device suggested by M(RT)^2. Transitions are *proposed* from, say, Y to X' using essentially *any* distribution $T(X' \mid Y)$. Then on comparing $f(X')$ with $f(Y)$ and taking into account T as well, the system is either moved to X' (move "accepted") or returned to Y (move "rejected"). Acceptance of the move occurs with probability $A(X' \mid Y)$, which must be calculated so as to satisfy detailed balance.

We then have

$$K(X \mid Y) = A(X \mid Y) T(X \mid Y). \tag{3.88}$$

Detailed balance requires

$$A(X \mid Y) T(X \mid Y) f(Y) = A(Y \mid X) T(Y \mid X) f(X). \tag{3.89}$$

We expect that the ratio

$$\frac{T(Y|X)f(X)}{T(X|Y)f(Y)}$$

will play a significant role in determining A.

Given a pdf $f(X)$, where X is a many-dimensional vector, the M(RT)^2 technique establishes a random walk whose steps are designed so that when repeated again and again, the asymptotic distribution of X's is $f(X)$. Suppose that $X_1, X_2, X_3, \ldots, X_n$ are the steps in a random walk. Each of the X's is a random variable and has an associated probability $\phi_1(X), \phi_2(X), \phi_3(X), \ldots, \phi_n(X)$, where $\phi_1(X)$ can be any distribution for X. The $\phi_n(X)$ have the property that asymptotically

$$\lim_{n\to\infty} \phi_n(X) = f(X).$$

At each step in the random walk there is a transition density $T(X \mid Y)$ that is the probability density function for a trial move to X from Y. The $T(X \mid Y)$ are normalized such that

$$\int T(X \mid Y)\, dX = 1$$

for all values of Y. A quantity $q(X \mid Y)$ is defined as

$$q(X \mid Y) = \frac{T(Y|X)f(X)}{T(X \mid Y)f(Y)} \geq 0, \tag{3.90}$$

where we explicitly assume that it is possible to move from X to Y if one can move from Y to X and vice versa. From $q(X \mid Y)$ the probability of accepting a move can be calculated; one frequently used possibility is

$$A(X \mid Y) = \min(1, q(X \mid Y)). \tag{3.91}$$

The algorithm can now be described concretely. At step n of the random walk, the value of X is X_n; a possible next value for X, X'_{n+1}, is sampled from $T(X'_{n+1} \mid X_n)$, and the probability of accepting X'_{n+1} is computed. If $q(X'_{n+1} \mid X_n) > 1$ then $A(X'_{n+1} \mid X_n) = 1$; if $q(X'_{n+1} \mid X_n) < 1$, then $A(X'_{n+1} \mid X_n) = q(X'_{n+1} \mid X_n)$, where

$$q(X'_{n+1} \mid X_n) = \frac{T(X_n \mid X'_{n+1})f(X'_{n+1})}{T(X'_{n+1} \mid X_n)f(X_n)}.$$

With probability $A(X'_{n+1} \mid X_n)$ we set $X_{n+1} = X'_{n+1}$; otherwise we set $X_{n+1} = X_n$. That is, if $A(X'_{n+1} \mid X_n) > \xi$, then $X_{n+1} = X'_{n+1}$; otherwise $X_{n+1} = X_n$. For $q(X'_{n+1} \mid X_n) > 1$, X_{n+1} will always equal X'_{n+1}. This procedure contains an element of rejection; however, if a X'_{n+1} is not accepted, we use the previous value rather than sample a new value.

As the random walk proceeds, a recursive relationship develops between succeeding $\phi_n(X)$'s. Let $\phi_n(X)$ be the distribution of values of X_n; what is the distribution ϕ_{n+1} for the values of X_{n+1}? There are two contributions to the distribution of the X_{n+1}: the probability of entering into the vicinity dX of X when we successfully move from X_n and the probability that once we are at X, we will stay at X. If we start out at some value Y contained in dY, the probability of moving from the neighborhood of Y to the neighborhood of X is $T(X \mid Y)\phi_n(Y)\,dY$. The probability of successfully moving from Y to X is $A(X \mid Y)T(X \mid Y)\phi_n(Y)\,dY$, so the net probability of successfully moving from any point Y to a neighborhood of X becomes

$$\int A(X \mid Y)T(X \mid Y)\phi_n(Y)\,dY. \tag{3.92}$$

In a similar manner, the net probability that a move away from X is not accepted is

$$\int (1 - A(Y \mid X))T(Y \mid X)\,dY, \tag{3.93}$$

where $T(Y \mid X)$ is the probability of moving from X to Y and $[1 - A(Y \mid X)]$ is the probability that the move was not accepted. Upon multiplying Eq. (3.93) by $\phi_n(X)$, the probability that we were at X, the relationship for $\phi_{n+1}(X)$ becomes

$$\phi_{n+1}(X) = \int A(X \mid Y)T(X \mid Y)\phi_n(Y)\,dY + \phi_n(X)\int [1 - A(Y \mid X)]T(Y \mid X)\,dY. \tag{3.94}$$

The random walk generates a recursion relationship for the density functions.

Earlier we asserted that the asymptotic distribution sampled in the random walk would be $f(X)$. According to a theorem in Feller,[11] if a random walk defines a system that is ergodic, then an asymptotic pdf exists and is unique if

$$\phi_n(X) = f(X) \Rightarrow \phi_{n+1} = f(x), \tag{3.95}$$

that is, if $f(x)$ is a stationary point of the recursion. Systems defined by random walks can be partitioned into several categories. If, in a random walk, the probability of returning to a neighborhood about X is 0, then the system is called a null system and the expected recurrence time is infinite. An example would be a one-dimensional system where X_{n+1} is constrained to be greater than X_n. A system where the random walk will return to the neighborhood of X every T steps is called periodic. An ergodic system is one in which the random walk may return to the neighborhood of X but does not do so periodically; it is neither null nor periodic.

The system generated by the M(RT)^2 sampling method is ergodic, but the proof will be omitted.[12] The system obeys detailed balance, which guarantees that if we can move from X to Y we can move from Y to X and the expected number of moves is the same in each direction when the asymptotic behavior is reached:

$$A(X \mid Y)T(X \mid Y)f(Y) = A(Y \mid X)T(Y \mid X)f(X). \tag{3.96}$$

The left-hand side is the net number of moves from Y to X and the right-hand side is the net number of moves from X to Y. Equation (3.96) is easily proved by using $q(X \mid Y)q(Y \mid X) = 1$, which follows from the definition of $q(X \mid Y)$. Suppose that $q(X \mid Y) > 1$; then $A(X \mid Y) = 1$. $A(Y \mid X) = q(Y \mid X)$ since $q(Y \mid X) < 1$. Substituting these into Eq. (3.96), we derive

$$\begin{aligned} T(X \mid Y)f(Y) &= q(Y \mid X)T(Y \mid X)f(X) \\ &= \frac{T(X \mid Y)f(Y)}{T(Y \mid X)f(X)}T(Y \mid X)f(X). \end{aligned}$$

The same answer would have resulted if we had chosen $q(Y \mid X) > 1$, so the algorithm developed in Eq. (3.90) and (3.91) satisfies detailed balance.

If we set $\phi_n(X) = f(X)$ in Eq. (3.94), the resulting equation is

$$\begin{aligned} \phi_{n+1}(X) &= \int A(X \mid Y)T(X \mid Y)f(Y)\,dY \\ &\quad + \int [1 - A(Y \mid X)]T(Y \mid X)f(X)\,dY. \end{aligned}$$

The first integral cancels the negative portion of the second integral by detailed balance, and we are left with

$$\begin{aligned}\phi_{n+1}(X) &= \int T(Y \mid X) f(X)\, dY \\ &= f(X)\end{aligned}$$

since $\int T(Y \mid X)\, dY = 1$. Therefore $f(X)$ is guaranteed to be the asymptotic distribution of the random walk.

The form for the probability of accepting a move is not limited to that given in Eq. (3.91). Another relation that has been used is

$$A'(X \mid Y) = \frac{q(X \mid Y)}{1 + q(X \mid Y)}$$

and (3.97)

$$A'(Y \mid X) = \frac{q(Y \mid X)}{1 + q(Y \mid X)} = \frac{1}{q(X \mid Y) + 1}.$$

A random walk whose probability of accepting a move is governed by (3.97) also exhibits detailed balance:

$$A'(X \mid Y) T(X \mid Y) f(Y) = A'(Y \mid X) T(Y \mid X) f(X).$$

Upon substituting Eq. (3.97) for A' on both sides of the equation we find

$$\frac{q(X \mid Y)}{1 + q(X \mid Y)} T(X \mid Y) f(Y) = \frac{1}{1 + q(X \mid Y)} T(Y \mid X) f(X), \quad (3.98)$$

which reduces to

$$\frac{T(Y \mid X) f(X)}{T(X \mid Y) f(Y)} T(X \mid Y) f(Y) = T(Y \mid X) f(X).$$

Either form for $A(X \mid Y)$, Eq. (3.91) or (3.97), may be used in the $M(RT)^2$ method. The former has been shown to give more rapid convergence in certain cases.[13]

In much of the literature of statistical physics, when a Monte Carlo calculation is mentioned what is meant is an application of the method of Metropolis et al. The great utility of the $M(RT)^2$ method is that it enables us to sample very complicated many-dimensional probability distribution functions in a simple, straightforward way. Unfortunately, the method does have major drawbacks of which the user must be aware. We are guaranteed to sample $f(X)$, but only asymptotically; therefore we must throw away L steps of the random walk until the steps are being sampled from $f(X)$. Furthermore, L is very difficult to estimate in advance. Normally substantial trial and error is used to estimate an appropriate value. The number of steps discarded may be minimized by selecting a $\phi_1(X)$ that is as close an approximation to $f(X)$ as possible. In addition, by making $T(X \mid Y)$ approximate $f(X)$, rapid convergence and small correlation are obtained. Note that were it possible to sample $T(X \mid Y) = f(X)$ exactly, then [cf. Eqs. (3.90) and (3.91)] all moves are accepted and the samples are independent. In that case, of course, one would not resort to the M(RT)^2 random walk, but it is clear that approximations to this limiting case may be fruitful.

In a Monte Carlo calculation, we are often trying to evaluate quantities of the form

$$G = \frac{\int g(X) f(X)\, dX}{\int f(X)\, dX}.$$

For example, if we were trying to simulate the equilibrium properties of a many-body system, G might be the energy or the *radial distribution function* (the probability that pairs of particles are found at varying separations). In practice, for each quantity of interest, a different number of steps in the random walk may have to be discarded since the asymptotic limit of the system is reached at varying rates. The averaging over the steps in the random walk begins only after the L steps have been thrown away, that is,

$$G = \frac{\int g(X) f(X)\, dX}{\int f(X)\, dX} = \sum_{n=L}^{L+N-1} \frac{g(X_n)}{N}.$$

The successive X's in the random walk are *not* independent and in most circumstances there is positive correlation. The variance of the calculated G will then be larger than if the steps were independent.

The number of steps to be discarded before averaging must be determined experimentally. One way is to average some property of the system, G, over (say) every 100 steps and observe the behavior of G with increasing walk length. An illustration of the behavior of G is shown in Figure 3.15. A decision is then made about where in the random walk the value of G has converged; further variation is just the normal

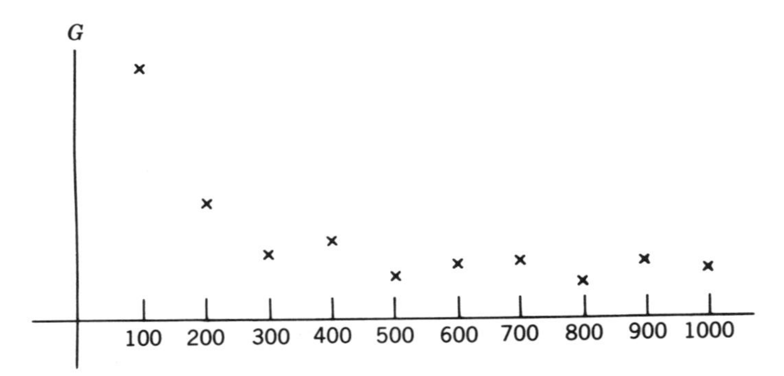

Figure 3.15. Behavior of $\langle G \rangle$ with length of the $\mathrm{M(RT)^2}$ random walk.

wandering that occurs in a random walk. All contributions prior to this point are discarded, and an average for G is extracted from the remaining steps. Careless observation of the attainment of the asymptotic distribution in the $\mathrm{M(RT)^2}$ method has led to some bad Monte Carlo calculation in the past.

The usage of the $\mathrm{M(RT)^2}$ method can be illustrated by some trivial examples. Almost without exception in applications, the transition density $T(X \mid Y)$ is assumed to be constant over a domain in X. Consider the one-dimensional pdf shown in Figure 3.16, defined on $(0, 1)$. A possible algorithm is to sample an x uniformly in a domain centered on Y and define the transition probability as

$$T(X \mid Y) = \begin{cases} \dfrac{1}{\Delta} & \text{if } |X - Y| < \dfrac{\Delta}{2} \\ 0 & \text{otherwise,} \end{cases}$$

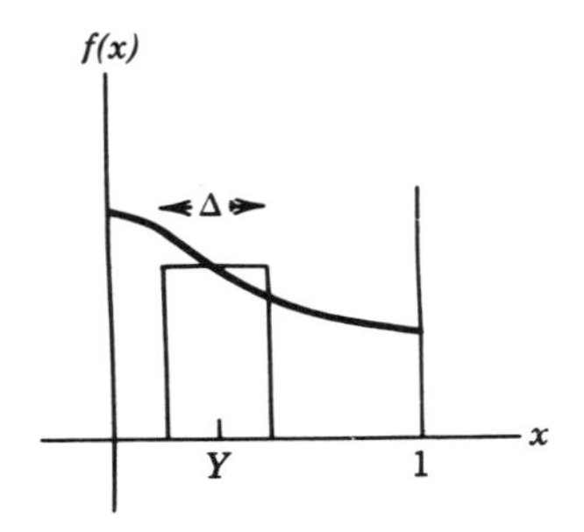

Figure 3.16. Monotonically decreasing pdf.

where Δ is the width of the domain about Y. For this choice of $T(X \mid Y)$, $q(X \mid Y)$ becomes

$$q(X \mid Y) = \frac{T(Y \mid X) f(X)}{T(X \mid Y) f(Y)} = \frac{(1/\Delta) f(X)}{(1/\Delta) f(Y)} = \frac{f(X)}{f(Y)}. \tag{3.99}$$

(Note that this is the usual method for continuous distribution f.) In this example detailed balance requires that if an interval about Y contains X, the corresponding interval about X must contain Y. For values of Y not in the vicinity of the origin, moves to the left will always be accepted, whereas moves to the right will be accepted with probability $q(X \mid Y) = f(X)/f(Y)$. When Y is in the vicinity of the origin, all moves to the left tend to be rejected since $f(X) = 0$ when $x < 0$.

An equally good algorithm for the same example would be to choose

$$T(X \mid Y) = \begin{cases} 1, & X \in (0, 1) \\ 0, & X \notin (0, 1). \end{cases}$$

This corresponds to a uniform random sampling of X in (0, 1).

The details of the M(RT)² method can be fully worked out for a one-dimensional example in which $f(X)$ is $2X$ on (0, 1) and is 0 elsewhere. We can choose the transition density to be

$$T(X \mid Y) = \begin{cases} 1, & X \in (0, 1) \\ 0, & \text{elsewhere} \end{cases}$$

and $q(X \mid Y)$ becomes, from Eq. (3.90),

$$q(X \mid Y) = \begin{cases} \dfrac{T(Y \mid X) f(X)}{T(X \mid Y) f(Y)} = \dfrac{X}{Y}, & X \in (0, 1) \\ 0, & \text{elsewhere.} \end{cases}$$

$q(X \mid Y) < 1$ implies that $f(X) < f(Y)$ and $X < Y$. This occurs for a move to the left, which will then be accepted with probability X/Y. When $q(X \mid Y) > 1$, $f(X) > f(Y)$ and a move to the right occurs. Such a move is always accepted in this example. The distribution of X's at the $(n+1)$th step will be, from Eq. (3.94),

$$\phi_{n+1}(X) = \int_X^1 \frac{X}{Y} \phi_n(Y)\, dY + \int_0^X \phi_n(Y)\, dY + \phi_n(X) \int_0^X \left(1 - \frac{Y}{X}\right) dY. \tag{3.100}$$

The first integral corresponds to the contribution from $q(X \mid Y) < 1$; the second contains the contributions when $q(X \mid Y) > 1$ (for which the probability of accepting the move is 1); and the third is the contribution from rejected steps. Normalization is preserved in Eq. (3.100). That is, if $\phi_n(X)$ is normalized, $\phi_{n+1}(X)$ will be too. If at some value of n, ϕ_n is linear and homogeneous, $\phi_n(X) = cX$, then $\phi_{n+1}(X) = cX$. This is true because Eq. (3.100) is homogeneous and because putting $\phi_n(X) = 2X$ must and does generate $\phi_{n+1}(X) = 2X$.

Let us assume that at the nth step the distribution of X's is

$$\phi_n(X) = a_n X + c_n X^{n+2} \tag{3.101}$$

and

$$\phi_0(X) = 3X^2.$$

Equation (3.101) is substituted into Eq. (3.100), the linear term $a_n X$ will carry through to give a contribution of $a_n X$ to $\phi_{n+1}(X)$. Applying the right-hand side of Eq. (3.100) to $c_n X^{n+2}$, we get

$$c_n X \int_X^1 \frac{Y^{n+2}}{Y}\, dY + c_n \int_0^X Y^{n+2}\, dY + c_n X^{n+2} \int_0^X \left(1 - \frac{Y}{X}\right) dY$$

$$= c_n \frac{X}{n+2}(1 - X^{n+2}) + \frac{c_n}{n+3} X^{n+3} + \frac{c_n}{2} X^{n+3},$$

which contains terms of both X and X^{n+3}. The form of $\phi_{n+1}(X)$ becomes

$$\phi_{n+1}(X) = \left(a_n + \frac{c_n}{n+2}\right) X + c_n \left[\frac{1}{2} - \frac{1}{(n+2)(n+3)}\right] X^{n+3}, \tag{3.102}$$

so asymptotically c_n will decrease as 2^{-n}, and we can rewrite $\phi_{n+1}(X)$:

$$\phi_{n+1}(X) \cong a_{n+1} X + \frac{c}{2^n} X^{n+3}.$$

The distributions of X exhibit explicit convergence; as n increases, the second term, $(c/2^n)X^{n+3}$, will contribute only values of X near 1 (see Figure 3.17). The normalization of the $\phi_n(X)$ gives the form of a_{n+1} as

$$a_{n+1} \cong 2 - \frac{c}{(n+4)2^n},$$

which asymptotically approaches 2.

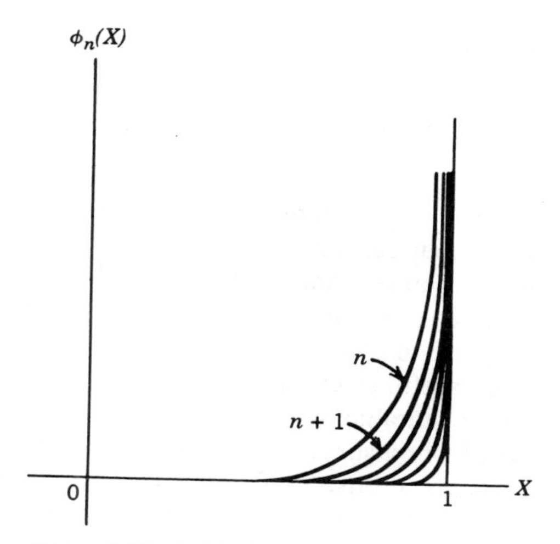

Figure 3.17. Behavior of $\phi_n(X)$ as n increases.

We have set $\phi_0(X) = 3X^2$, so $a_0 = 0$ and $c_0 = 3$. For $n+1 = 1$, Eq. (3.102) yields

$$\phi_1(X) = \tfrac{3}{2}X + X^3$$

or that $a_1 = \frac{3}{2}$ and $c_1 = 1$. Another iteration of Eq. (3.102) gives

$$\phi_2(X) = \tfrac{11}{6}X + \tfrac{5}{12}X^4,$$

where $a_2 = \frac{11}{6}$ and $c_2 = \frac{5}{12}$; even after only two steps a_n is easily seen to be approaching 2 and c_n is decreasing as approximately 2^{-n}.

The fact that at the nth stage $\phi_n(X)$ differs from a_nX by a function like X^{n+2} shows that the error in integrating $\int \phi_n(X)g(X)\, dX / \int \phi_n\, dX$ instead of $\int 2Xg(X)\, dX$ will depend on $g(X)$. This is a special case of the remark made earlier that the number of steps to be discarded may depend on the properties of a system being computed.

3.8. APPLICATION OF M(RT)²

As we have seen, when a pdf $f(x)$ must be sampled, M(RT)² gives a prescription for defining a random walk $X_0, X_1, \ldots$ for which the asymptotic distribution is guaranteed to be $f(X)$. These points X_i may be used to evaluate some integral by setting

$$\int g(X) f(X)\, dX \cong \frac{1}{M} \sum g(X_i).$$

Because $f(X)$ is sampled only asymptotically, the estimate of the integral is biased by an amount that can be made smaller by discarding more and more X_i at the beginning of the walk and by extending the walk to larger M. As we have indicated, the question of how many must be discarded can be answered only in the context of a specific problem. Nevertheless, it is necessary to give careful attention to this problem. It is this difficulty that makes $\mathrm{M(RT)^2}$ unsuitable for applications in which only a few variables are sampled from a specific $f(X)$ with fixed parameters.

In the most usual applications, $T(X \mid Y)$ is taken to be uniform over a small domain (e.g., a square or cube of side s) in the space of X or in a subspace thereof. Let us consider a highly simplified model of atoms moving on the surface of something. Imagine a two-dimensional square box of side L that contains a large number (10^2–10^4) of disks (Figure 3.18). Each disk is located by the coordinates of its center and has an effective diameter a. If there are M disks, then we have $2M$ position coordinates, and a configuration of the system can be indicated by the vector

$$\mathbf{X} = (x_1, y_1, x_2, y_2, \ldots, x_m, y_m).$$

The pdf describing the system, $f(\mathbf{X})$, is a constant except when $x_k < 0$, $x_k > L$, or $(x_l - x_k)^2 + (y_l - y_k)^2 < a^2$ for any l and k. If any one of these conditions happens, $f(\mathbf{X}) = 0$. These restrictions prevent the disks from moving outside the box and from overlapping each other.

Given the two-dimensional system detailed above, what is the prob-

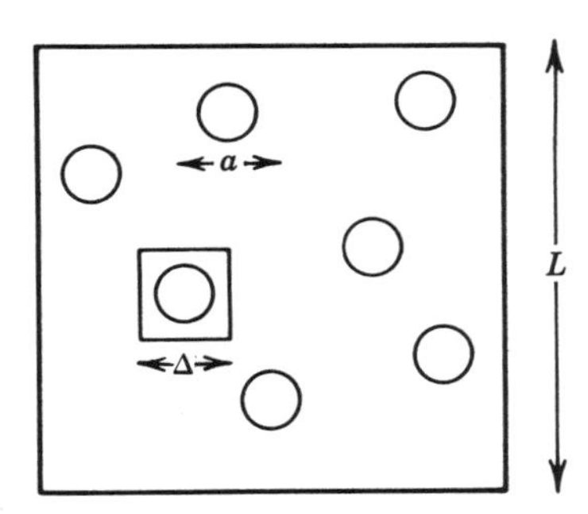

Figure 3.18. Hard spheres in a box.

ability that two disks will be a distance r apart? We can answer this problem by using the M(RT)^2 method to generate many configurations $\mathbf{X}$ and measure the frequency of occurrence of r. The problem would be difficult to analyze by any method other than M(RT)^2. A scheme for analyzing the problem in this way is to start with the disks on a lattice in the box so as to guarantee no overlap at the start. A disk is chosen randomly to be moved; the transition probability for moving from one configuration to another (the configurations differ by moving one disk) is

$$T(\mathbf{X} \mid \mathbf{Y}) = \sum f_l t(x_l, y_l \mid x'_l, y'_l),$$

where

$$f_l = 1/M$$

and $t(x_l, y_l \mid x'_l, y'_l)$ is the transition density for moving uniformly in the little square. It is simply a constant, $1/\Delta^2$ where Δ is the side of the little square as in Figure 3.18. Thus, T is a constant and the ratio $T(\mathbf{Y} \mid \mathbf{X})/T(\mathbf{X} \mid \mathbf{Y})$ that appears in the acceptance probability is exactly 1. Once x_l and y_l have been calculated, the ratio $f(\mathbf{X})/f(\mathbf{Y})$ is evaluated. If the step has moved a disk outside the box or some disks are overlapping, then $f(\mathbf{X}) = 0$ and the move is rejected; otherwise $f(\mathbf{X})/f(\mathbf{Y}) = 1$ and the move is accepted. The process is repeated many times to generate many different configurations and a frequency function for r is tabulated. The moves that occur do not describe physical motion of the disks since the "kinetics" introduced is completely artificial.

Some care must be exerted in the selection of the size of the little box surrounding a disk. In the limit that the individual box is the size of the whole domain (the big box) it will be unlikely to find an empty space in which to move a disk. Almost all the moves will be rejected, so the random walk will be composed of the same configurations repeated many times. In this case, the sequential correlation from step to step is very close to one and the random walk provides very little new information at each step.

At the other extreme is a box too small. The move of the disk is always accepted, but the configuration changes very slowly. Again, there is high sequential correlation with little new information emerging as the particles are moved.

The "lore" for using the M(RT)^2 method recommends that the average acceptance probability should be approximately 50% to avoid both extremes described above. That is, the box size should be chosen such that 50% of the moves within it are accepted. A criterion for choosing the box size that has better justification but is harder to apply, is to make moves within a box so that the variance in the desired result is minimized for given computing time. In this case, the variance is minimized by reducing the sequential correlation between steps. Sequential correlation decreases as the mean-square displacement increases. The mean-square displacement in one step is proportional to $p_A\Delta^2$. A simple usable criterion is to maximize this quantity where p_A is the average probability of acceptance. An acceptable value for Δ can be established by several trial runs in which Δ is changed and the variation in $\langle r^2 \rangle$ and p_A observed. The value of Δ that produces the maximum $\langle r^2 \rangle$ is then used in the actual calculation. This criterion has proved useful in practical cases. More generally one should take account of the computer time used in the computation of $f(\mathbf{X})$.

3.9. TESTING SAMPLING METHODS

It is important to test that an algorithm indeed samples $f(x)$. Several methods to check the algorithm can be used. For few-dimensional distributions one may generate random variables and sort into bins within the range of the random variable. A frequency function is yielded by this method, and elementary statistical tests may be applied to it, for example, the number of times the observed frequency function is higher or lower than the expected value. A chi-squared test is appropriate. A third way of checking an algorithm is to use the algorithm to evaluate an elementary integral whose value is known, $\int g(x)f(x)\,dx$, and decide whether the value agrees within statistics with the correct value. Moments, in which $g_n = x^n$, are especially useful.

Consider the algorithm proposed by Wiener to sample two independent gaussian random variables given by Eq. (3.38). A subroutine is easily written to generate the random variables; we test this code by sampling from it many times and binning the result. After 5000 random samples our frequency function for $\phi(x \mid 0, 1)$ appears as the bar graph* in Figure 3.19. The solid line represents the analytic values of Eq. (3.36).

A combination of such tests along with analysis of the hypothesis that the population is drawn from $f(x)$ is often warranted.

Such tests, if carefully carried out, constitute useful confirmations of

*As with any random sampling experiment, the details of the structure of this bar graph depend on the sequence of pseudo-random numbers used. A different sequence will give a different graph.

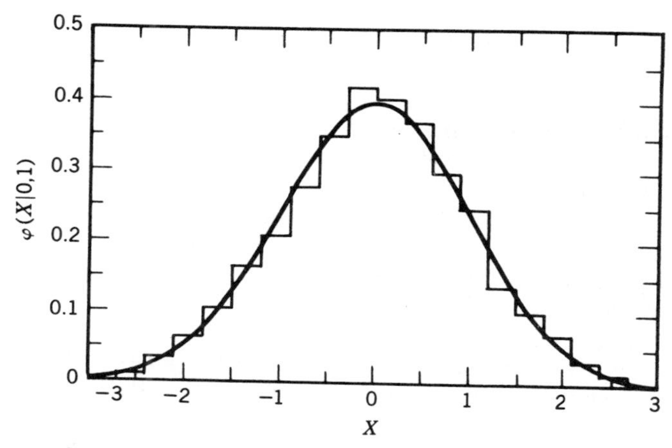

Figure 3.19. Frequency function after 1000 random samples of $\phi(x \mid 0, 1)$.

the quality of the pseudo-random number sequences as well as of the sampling method under study.

The combination of logical complexity and statistical fluctuation mean that Monte Carlo programs may require care in testing. Proper modular design of such programs is essential, and sampling routines constitute appropriate modules. The sampling routines should be thoroughly tested with specially designed "drivers," which evaluate the performance of the routines over the range of expected parameters. The final Monte Carlo code should be assembled from well-tested modules. Completion and analysis of a significant computation is not the ideal stage at which to uncover a "bug" in a sampling routine.

REFERENCES

1. H. Kahn, Applications of Monte Carlo, AECU-3259, 1954; C. J. Everett and E. D. Cashwell, A Third Monte Carlo Sampler, Los Alamos Technical Report LA-9721-MS, 1983.
2. G. E. P. Box and M. E. Muller, A note on the generation of random normal deviates, *Ann. Math. Statist.*, **29**, 610, 1958.
3. J. von Neumann, Various techniques used in connection with random digits, National Bureau of Standards, Applied Math. Series, Vol. 12, 36, 1951.
4. G. Marsaglia, M. D. MacLaren, and T. A. Bray, A fast procedure for generating random variables, *Comm. ACM*, **7**, 4, 1964.

5. F. B. Brown, W. R. Martin, and D. A. Calahan, A discrete sampling method of vectorized Monte Carlo calculations, *Trans. Am. Nuc. Soc.*, **38**, 354, 1981.
6. N. Metropolis, A. W. Rosenbluth, M. N. Rosenbluth, A. H. Teller, and E. Teller, Equations of state calculations by fast computing machines, *J. Chem. Phys.*, **21**, 1087, 1953.
7. J. P. Valleau and S. G. Whittington, A Guide to Monte Carlo for Statistical Mechanics: 1. Highways, in *Statistical Mechanics, Part A: Equilibrium Techniques*, Modern Theoretical Chemistry Series, Vol. 5, B. Berne, Ed., Plenum, New York, 1976, Chap. 4.
8. J. P. Valleau and G. M. Torrie, A Guide to Monte Carlo for Statistical Mechanics: 2. Byways, in *Statistical Mechanics, Part A: Equilibrium Techniques*, Modern Theoretical Chemistry Series, Vol. 5, B. Berne, Ed., Plenum, New York, 1976, Chap. 5.
9. W. W. Wood and J. J. Erpenbeck, Molecular dynamics and Monte Carlo calculations in statistical mechanics, *Ann. Rev. Phys. Chem.*, **27**, 319, 1976.
10. W. W. Wood, Monte Carlo studies of simple liquid models, in *The Physics of Simple Liquids*, H. N. V. Temperley, J. S. Rowlinson, and G. S. Rushbrooke, Eds., North-Holland, Amsterdam, 1968, Chap. 5, Sect. 1.
11. W. Feller, *An Introduction to Probability Theory and Its Applications*, Vol. 1, John Wiley and Sons, New York, 1950, p. 324.
12. W. W. Wood and F. R. Parker, Monte Carlo equation of state of molecules interacting with the Lennard-Jones potential. I. Supercritical isotherm at about twice the critical temperature, *J. Chem. Phys.*, **27**, 720, 1957.
13. J. P. Valleau and S. G. Whittington, *op. cit.*, Chap. 4.

GENERAL REFERENCES

C. J. Everett and E. D. Cashwell, A Third Monte Carlo Sampler, Los Alamos Technical Report LA-9721-MS, 1983.

H. Kahn, Applications of Monte Carlo, AECU-3259, 1954.

R. Y. Rubinstein, *Simulation and the Monte Carlo Method*, John Wiley and Sons, New York, 1981, Chap. 3 and references therein.

4 MONTE CARLO EVALUATION OF FINITE-DIMENSIONAL INTEGRALS

In this chapter we explore somewhat more systematically the ideas that underlie Monte Carlo quadrature. If an integral must be evaluated having the form

$$G = \int_{\Omega_0} g(X) f(X)\, dX, \tag{4.1}$$

where

$$f(X) \geq 0, \qquad \int_{\Omega_0} f(X)\, dX = 1, \tag{4.2}$$

then the following *game of chance* may be used to make numerical estimates. We draw a set of variables $X_1, X_2, \ldots, X_N$ from $f(X)$ [i.e., we "sample" the probability distribution function $f(X)$ in the sense defined in Chapter 3] and form the arithmetic mean

$$G_N = \frac{1}{N} \sum_i g(X_i). \tag{4.3}$$

The quantity G_N is an estimator for G and the fundamental theorem of Monte Carlo guarantees that

$$\langle G_N \rangle = G$$

if the integral [Eq. (4.1)] exists. Since G_N estimates G we can write

$$G_N = G + \text{error}.$$

If the variance exists, the error appearing in the last statement is a random variable whose mean is 0 and whose width is characterized for large N by

$$|\text{error}| = \epsilon \cong \frac{\sigma_1}{N^{1/2}},$$

where

$$\sigma_1^2 = \int g^2(X) f(X)\, dX - G^2. \tag{4.4}$$

The error estimate may be inverted to show the number of samples needed to yield a desired error, ϵ:

$$N = \sigma_1^2/\epsilon^2. \tag{4.5}$$

The integral to be done need not exhibit explicitly a function $f(X)$ satisfying the properties expressed by Eq. (4.2). One can simply use $f(X) = 1/\Omega_0$ and $g(X) = \Omega_0 \times \text{integrand}$. Later [cf. Eq. (4.10) et seq.] we shall discuss more general ways of introducing a distribution function.

The integral in Eq. (4.1) could also be evaluated by quadrature. Let us assume that the domain of integration is an n-dimensional unit hypercube; then a numerical integration procedure can be written:

$$G \cong \sum w_i g(X_i) f(X_i),$$

where X_i is a lattice of points that fills the unit hypercube and w_i is a series of quadrature weights. The error associated with this quadrature is bounded by

$$\epsilon \leq ch^k, \tag{4.6}$$

where h measures the size of the interval separating the individual X_i. The constants c and k depend on the actual numerical integration method used, and k normally increases with more accurate rules. The bound on the error in Eq. (4.6) is not a statistical variable, but is an absolute number. The actual error, however, is usually predictable to some degree.

If we assume that the time necessary for a computation will be proportional to the total number of points used, then

$$T_c \propto N = N_0\left(\frac{1}{h}\right)^n, \tag{4.7}$$

where N_0 is a constant of the order of 1 and n is the number of dimensions. Equation (4.6) can be rewritten

$$h \geq \left(\frac{\epsilon}{c}\right)^{1/k},$$

and Eq. (4.7) becomes

$$\begin{aligned} T_c &\propto N_0\left(\frac{c}{\epsilon}\right)^{n/k} \\ &= t_0\epsilon^{-n/k}. \end{aligned} \tag{4.8}$$

The greater the accuracy demanded in a calculation, the greater the computational time will be.

In a Monte Carlo calculation, the total computation time is the product of the time for an individual sampling of X, t_1, times the total number of points;

$$T_c = t_1 N.$$

From Eq. (4.5) this may be rewritten

$$T_c = t_1\sigma_1^2/\epsilon^2 = t_1'\epsilon^{-2};$$

the exponent of ϵ is the same in any number of dimensions. For large n, it is difficult to find a k in Eqs. (4.6) and (4.8) such that $n/k < 2$, so asymptotically ($n \to \infty$) a Monte Carlo calculation is more advantageous than a numerical integration of Eq. (4.1). The Monte Carlo calculation will take less total time for the same value of ϵ. This assumes that the two error estimates can be directly compared.

In spite of the apparently slow convergence ($\cong N^{-1/2}$) of the error of Monte Carlo quadrature, it is in fact more efficient computationally than finite difference methods in dimensions higher than six to ten.

Two different Monte Carlo evaluations of an integral can have differing variances. The quantity

$$Q_1 = t_1\sigma_1^2 \tag{4.9}$$

is a measure of the quality (efficiency) of a Monte Carlo calculation. The decision on which Monte Carlo algorithm to use in a large computation can be based on the values of Q_1 extracted from some trial calculations. A common phenomenon is for t to increase as σ decreases through a more elaborate Monte Carlo algorithm. The question is then whether the decrease in σ^2 will more than compensate for the increase in time. It will if Q decreases.

Three major classes of techniques are used to reduce the variance in Monte Carlo quadrature.

1. Importance sampling can be introduced into the calculation to increase the likelihood of sampling variables where the function is large or rapidly varying.
2. The expected value of a random variable can be used rather than the variable itself. This substitution never increases variance and many times will substantially reduce it.
3. Correlations between succeeding samples may be exploited to advantage. In *control variates*, an easily evaluated approximation to the integrand is used to reduce the variance. If successive random variables are negatively correlated, the variance will be smaller than if they were independent. The technique called *antithetic variates* exploits the reduction in variance that results when negatively correlated samples are deliberately produced and grouped together.

4.1. IMPORTANCE SAMPLING

Suppose we have an n-dimensional integral

$$G = \int g(X) f(X)\, dX$$

that we wish to evaluate. The function $f(X)$ is not necessarily the best pdf to use in the Monte Carlo calculation even though it appears in the integrand. A different pdf, $\tilde{f}(X)$, can be introduced into the integral as follows:

$$G = \int \left[\frac{g(X) f(X)}{\tilde{f}(X)}\right] \tilde{f}(X)\, dX, \tag{4.10}$$

where

$$\tilde{f}(x) \geq 0, \qquad \int \tilde{f}(X)\, dX = 1, \tag{4.11}$$

and $g(X)f(X)/\tilde{f}(X) < \infty$ except perhaps on a (countable) set of points. The variance of G when $\tilde{f}(X)$ is used becomes

$$\text{var}\{G\}_{\tilde{f}} = \int \left[\frac{g^2(X)f^2(X)}{\tilde{f}^2(X)}\right] \tilde{f}(X)\,dX - G^2. \tag{4.12}$$

G^2 being fixed, we want the $\tilde{f}(X)$ that will minimize the quantity $\int [g^2(X)f^2(X)/\tilde{f}(X)]\,dX$. Of course, the integral is minimized by choosing $\tilde{f}(X)$ as large as we like, but we have the additional constraint expressed by Eq. (4.11). The function $\tilde{f}(X)$ that satisfies the criteria given above may be deduced by using a Lagrange multiplier λ. In this method we wish to find $\tilde{f}(X)$ such that

$$L\{\tilde{f}\} = \left[\int \frac{g^2(X)f^2(X)}{\tilde{f}(X)}\,dX + \lambda \int \tilde{f}(X)\,dX\right] \tag{4.13}$$

is minimized. We consider small variations of $\tilde{f}(X)$ on the quantity in brackets and set the variation of the quantity in brackets equal to zero

$$\frac{\delta}{\delta \tilde{f}}[\cdots] = 0.$$

Performing the functional differentiation yields

$$-\frac{g^2(X)f^2(X)}{\tilde{f}^2(X)} + \lambda = 0 \tag{4.14}$$

or

$$\tilde{f} = \lambda |g(X)f(X)|. \tag{4.15}$$

If the function $g(X)f(X)$ varied with X as the solid line in Figure 4.1, then $\tilde{f}(X)$ would be proportional to the dotted line. The value of λ may be found by requiring that $\int \tilde{f}(X)\,dX = 1$. If $g(X) \geq 0$, then $\tilde{f}(X) = \lambda g(X)f(X)$ and $\lambda = 1/G$, so

$$\tilde{f}(X) = \frac{g(X)f(X)}{G}. \tag{4.16}$$

A Monte Carlo algorithm to evaluate the integral would be to sample a

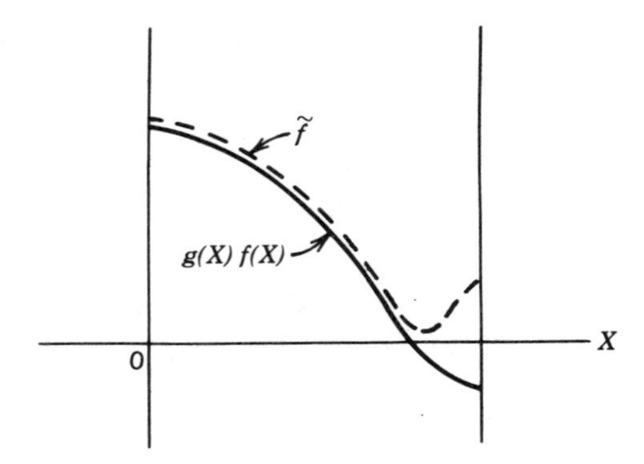

Figure 4.1. A function $g(X)f(X)$ and a possible $\tilde{f}$.

series of X_i from $\tilde{f}(X)$ and construct the sum

$$\tilde{G}_N = \frac{1}{N}\sum_{i=1}^{N}\frac{g(X_i)f(X_i)}{\tilde{f}(X_i)} = \frac{1}{N}\sum\frac{g(X_i)f(X_i)}{g(X_i)f(X_i)/G}$$
$$= \frac{1}{N}\sum_{1}^{N} G = G.$$

If we already know the correct answer G, the Monte Carlo calculation will certainly give it back with zero variance! This clearly corresponds to the minimum variance calculation. Although we cannot in practice use the $\tilde{f}(X)$ prescribed by Eq. (4.16), we expect that "similar" functions will reduce the variance. What is meant will be explored in some examples. One important criterion is that $g(X_i)f(X_i)/\tilde{f}(X_i)$ be bounded from above.

As the first example consider the integral

$$G = \int_0^1 \cos\left(\frac{\pi x}{2}\right) dx.$$

A straightforward Monte Carlo algorithm would be to sample x uniformly on (0, 1), ($f_1(x) = 1$), and to sum the quantity

$$g_1(x) = \cos\left(\frac{\pi x}{2}\right).$$

The variance of the population for a single sample of x may be analytically evaluated [Eq. (4.12)] and is

$$\text{var}\{g_1\} = 0.0947\ldots.$$

By expanding $\cos(\pi x/2)$ in a power series, a better choice for the importance function may be found,

$$\cos\left(\frac{\pi x}{2}\right) = 1 - \frac{\pi^2 x^2}{8} + \frac{\pi^4 x^4}{2^4 4!} - \cdots$$

and we can let

$$\tilde{f}(x) = \tfrac{3}{2}(1 - x^2).$$

This pdf looks like the integrand for small values of x.* The estimator for G is now

$$\tilde{g} = \frac{g_1}{\tilde{f}} = \frac{2}{3}\frac{\cos(\pi x/2)}{1 - x^2}$$

and the variance associated with a single sample is

$$\text{var}\{\tilde{g}\} = 0.000990.$$

With this choice of importance function the variance decreased by a factor of 100.

As another example consider the integral

$$\int_0^1 \sqrt{1 - x^2}\, dx = \frac{\pi}{4};$$

again the straightforward Monte Carlo method would be to sample x uniformly on (0, 1) and form the estimator

$$g_1 = \sqrt{1 - x^2}.$$

The variance associated with this procedure is

$$\text{var}\{g_1\} = 0.050.$$

*It can be sampled by generating ξ_1 and ξ_2. If $\xi_2 \le \xi_1(3 - \xi_1)/2$, set $x = 1 - \xi_1$; otherwise set $x = \frac{1}{2}((9 - 8\xi_2)^{1/2} - 1)$.

To improve the calculation, the integrand can be expanded in a power series about its maximum:

$$g = 1 - \tfrac{1}{2}x^2 + \cdots,$$

from which we infer that a reasonable importance function might be

$$\tilde{f}(x) = \frac{1 - \beta x^2}{1 - \frac{1}{3}\beta}. \tag{4.17}$$

The value of β may be chosen to give the minimum variance. With the choice of $\tilde{f}(x)$ given above, $\tilde{g}$ becomes

$$\tilde{g} = \frac{g_1}{\tilde{f}} = (1 - \tfrac{1}{3}\beta)\frac{\sqrt{1 - x^2}}{(1 - \beta x^2)}$$

and its variance is

$$\text{var}\{\tilde{g}\} = (1 - \tfrac{1}{3}\beta)\left[\frac{1}{\beta} - \frac{(1-\beta)}{\beta\sqrt{\beta}}\tanh^{-1}\sqrt{\beta}\right] - \left(\frac{\pi}{4}\right)^2.$$

Numerically minimizing the variance with respect to β gives var$\{g\} =$ 0.0029 when $\beta = 0.74$. If we chose β to be $\frac{1}{2}$ as implied by the power series, the minimum variance is not achieved (var$\{g\} = 0.011$), but is still substantially lower than the variance associated with straightforward Monte Carlo. Very simple importance sampling schemes and ad hoc choices of parameters can lower the variance dramatically.

An experimental method of obtaining a parameter like β in an importance function is to try the Monte Carlo calculation with different values of β and estimate the value that minimizes the variance of $\tilde{g}$. The optimal value for β is not located precisely by this process because the fluctuations mask the trend, but we can hope that the variance of the Monte Carlo calculation is approximately constant in a neighborhood of the optimal β. A sample search for a good value of β is shown in Figure 4.2.

In a global sense we want an importance function that matches the general behavior of the integrand at its maximum but is also similar over the whole range of integration. In Figure 4.3, g is the integrand we wish to approximate. $\tilde{f}_1$ is an importance function derived from a Taylor series expansion; it is always greater than g. The function $\tilde{f}_2$ is a better importance function because it describes g better over the entire range

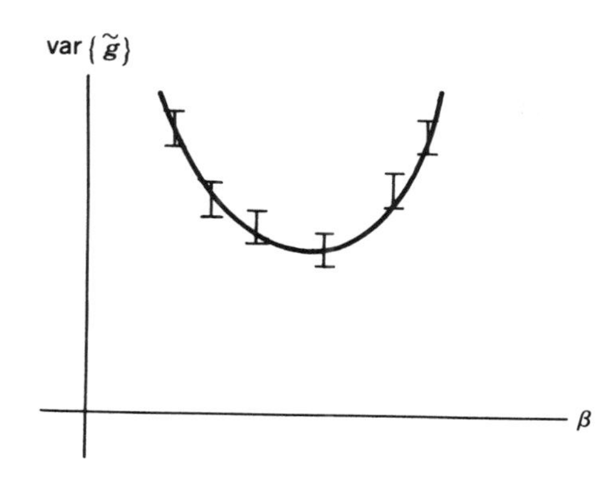

Figure 4.2. Finding the value of β that minimizes var{$\bar{g}$} graphically.

of integration. In the previous example, the parameter β in Eq. (4.17) could be chosen by letting

$$1 - \beta x^2 = \sqrt{1 - x^2}$$

when $\sqrt{1-x^2} = \frac{1}{2}$. The value of $\beta = \frac{2}{3}$ and the resulting variance of g is 0.0039, which is not much larger than the minimum variance. Therefore it is not always necessary to calculate the variance as a function of β to select a β that substantially reduces the variance. Simple arguments of a dimensional type can locate the optimal β.

When the integrand $g(X)f(X)$ of an integral is singular, var$\{g(X)\}$ may not exist. In this case we can always choose $\tilde{f}(X)$ such that the ratio $g(X)f(X)/\tilde{f}(X)$ is bounded. We assume here that $g(X)$ and $f(X)$ are known analytical functions whose singularities are easily identified. Consider the integral

$$\int \frac{dx}{x^{1/2}};$$

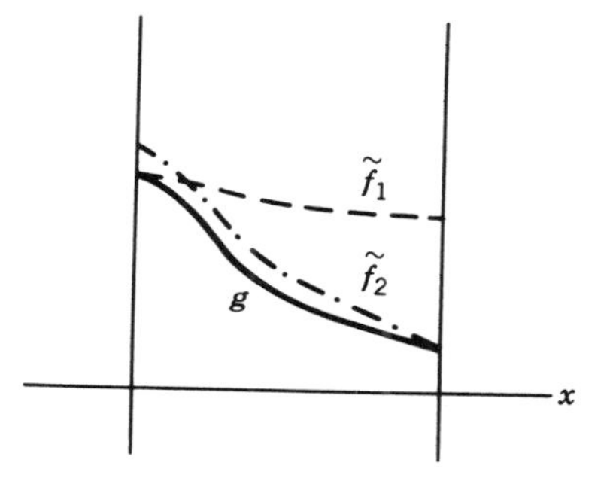

Figure 4.3. A function g and two possible importance functions $\tilde{f}_1$ and $\tilde{f}_2$.

the straightforward Monte Carlo calculation would be to sample x uniformly on (0, 1) with the estimator $g_1 = 1/x^{1/2}$. The variance contains $\langle g^2 \rangle$, which is

$$\langle g^2 \rangle = \int_0^1 \frac{dx}{x} = \infty$$

so that the variance for this calculation does not exist. As an alternative we can try

$$\tilde{f}(x) = (1-r)x^{-r},$$

where $r < 1$. The estimator for G is now

$$\tilde{g} = \frac{x^{r-1/2}}{1-r},$$

and the nth moment of $\tilde{g}$ is

$$\langle \tilde{g}^n \rangle = (1-r)^{-n+1} \int_0^1 x^{nr-n/2} x^{-r}\, dx.$$

For the integrals to exist,

$$(n-1)r - n/2 > -1,$$

and in particular, all moments will exist if $1 > r \geq \frac{1}{2}$. Of course, the optimal r is $r = \frac{1}{2}$.

As a second example, suppose we wish to evaluate

$$\int_0^1 \frac{dx}{[x(1-x)]^{1/2}};$$

again, a straightforward Monte Carlo calculation will have an infinite variance. To eliminate this, we shall introduce an $\tilde{f}(x)$ that has singularities both at 0 and at 1:

$$\tilde{f}(x) = \frac{1}{4\sqrt{x}} + \frac{1}{4\sqrt{1-x}}.$$

The method for sampling $\tilde{f}(x)$ was described in Section 3.4.4. The estimator $\tilde{g}(x)$ becomes

$$\tilde{g}(x) = \frac{\dfrac{4}{\sqrt{x(1-x)}}}{\dfrac{1}{\sqrt{x}} + \dfrac{1}{\sqrt{1-x}}} = \frac{4}{\sqrt{x} + \sqrt{(1-x)}}.$$

$\tilde{g}(x)$ is bounded by 4 from above and is greater than $4/\sqrt{2}$ (Figure 4.4). The variance associated with $\tilde{g}(x)$ is easily bounded:

$$\text{var}\{\tilde{g}\} = \int \tilde{f}(x)(\tilde{g} - G)^2\, dx < \left(4 - \frac{4}{\sqrt{2}}\right)^2 = 1.37,$$

and we have eliminated the infinite variance.

The strategy needed to develop an importance function is to identify the singularities, i.e.

$$g(x) \cong \begin{cases} x^{-1/2} & \text{near} \quad x = 0 \\ (1-x)^{-1/2} & \text{near} \quad x = 1, \end{cases}$$

and simply take a sum of the singularities as the importance function. From the composition method (Section 3.4) we know we can sample the sum. It is not really necessary to discuss exotic central limit theorems in Monte Carlo calculations, since we can always eliminate infinite variances through importance sampling.

As a last example we shall evaluate a singular integral of some practical interest. Let $\mathbf{X}$ and $\mathbf{X}_0$ be three-dimensional vectors contained

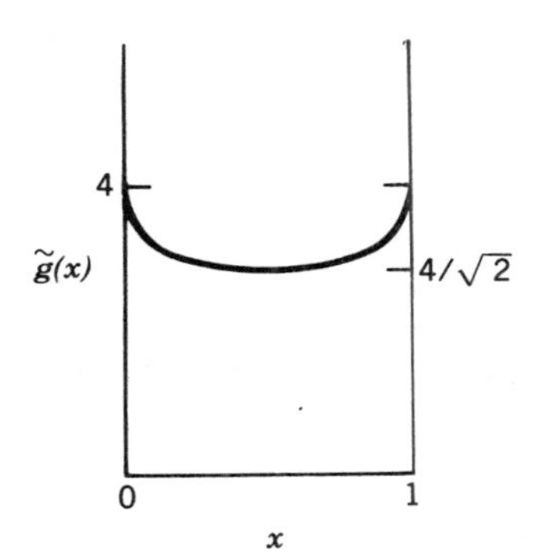

Figure 4.4. The estimator $\tilde{g}(x)$ as a function of x.

in the volume V and $S(\mathbf{X})$ a function bounded away from zero and infinity, $0 < S_0 \leq S(\mathbf{X}) < S_1 < \infty$. Consider the integral

$$\langle S \rangle = \int_V \frac{S(\mathbf{X}) \exp[-\mu|\mathbf{X}-\mathbf{X}_0|]}{|\mathbf{X}-\mathbf{X}_0|^2} \, d\mathbf{X}, \tag{4.18}$$

where

$$\int_V S(\mathbf{X}) \, d\mathbf{X} = 1.$$

Physically, this integral occurs in the neutral radiation transport problem with source $S(\mathbf{X})$, attenuation with coefficient μ, and an infinitesimal detector at $\mathbf{X}_0$. The "natural" way to evaluate Eq. (4.18) is to sample $\mathbf{X}$ from $S(\mathbf{X})$ and sum the quantity

$$g(\mathbf{X}) = \frac{\exp[-\mu|\mathbf{X}-\mathbf{X}_0|]}{|\mathbf{X}-\mathbf{X}_0|^2}. \tag{4.19}$$

We make a change of the coordinates by letting $\mathbf{X}-\mathbf{X}_0 = \mathbf{r}$. $d\mathbf{X} = d\mathbf{r}$ since $\mathbf{X}_0$ is fixed. The integral then becomes

$$\langle S \rangle = \int_V S(\mathbf{r}+\mathbf{X}_0) \frac{\exp[-\mu r]}{r^2} \, d\mathbf{r}. \tag{4.20}$$

The variance associated with the estimator of the integral in Eq. (4.19) will contain the quantity

$$\langle g^2 \rangle = \int_V S(\mathbf{r}+\mathbf{X}_0) \frac{\exp[-\mu r]}{r^4} \, d\mathbf{r};$$

we may bound $\langle g^2 \rangle$ by invoking the integral theorem of the mean. Since $S(\mathbf{X})$ is bounded away from 0,

$$\langle g^2 \rangle \geq S_0 \int_V \frac{\exp[-2\mu r]}{r^4} \, d\mathbf{r};$$

this is less than the integral over the largest sphere wholly contained in V

$$\langle g^2 \rangle \geq 4\pi S_0 \int \frac{\exp[-2\mu r]}{r^4} r^2 \, dr, \tag{4.21}$$

which diverges. Therefore the variance does not exist for the estimator in Eq. (4.19).

To have a finite-variance Monte Carlo calculation, we must change the method of sampling. Suppose that b is the radius of the smallest sphere centered at $\mathbf{X}_0$ that contains V; we can introduce a pdf, $f(\mathbf{r})$, for sampling within the sphere as

$$f(\mathbf{r}) = \frac{\mu}{4\pi(1-\exp[-\mu b])} \frac{\exp[-\mu r]}{r^2}. \tag{4.22}$$

To sample an $\mathbf{r}$ we write $f(\mathbf{r})\, d\mathbf{r}$ as

$$\begin{aligned} f(\mathbf{r})\, d\mathbf{r} &= f(r) r^2\, dr \sin\theta\, d\theta\, d\phi \\ &= \frac{\mu \exp[-\mu r]}{(1-\exp[-\mu b])}\, dr \frac{\sin\theta\, d\theta}{2} \frac{d\phi}{2\pi}. \end{aligned} \tag{4.23}$$

Sample $\cos\theta$ uniformly on $(-1, 1)$ and ϕ uniformly on $(0, 2\pi)$. If b were infinite, the r function would be $\mu \exp[-\mu r]\, dr$, which can be sampled by $r = -(1/\mu) \log \xi$. For finite b, $\mu \exp[-\mu r]/(1-\exp[-\mu b])$ may be sampled by setting

$$r = -\frac{1}{\mu} \log \xi \quad (\text{mod } b). \tag{4.24}$$

Using $f(\mathbf{r})$ in Eq. (4.22), the integral in Eq. (4.18) becomes

$$\langle S \rangle = \frac{(1-\exp[-\mu b]) 4\pi}{\mu} \int_V S(\mathbf{r}+\mathbf{X}_0) f(\mathbf{r})\, d\mathbf{r}$$

and the estimator for $\langle S \rangle$ is

$$\tilde{g}(\mathbf{r}) = \frac{(1-\exp[-\mu b]) 4\pi}{\mu} S(\mathbf{r}+\mathbf{X}_0).$$

The variance contains $\langle \tilde{g}^2 \rangle$,

$$\langle \tilde{g}^2 \rangle = \left(\frac{(1-\exp[-\mu b])}{\mu} 4\pi \right)^2 \int S^2(\mathbf{r}+\mathbf{X}_0) f(\mathbf{r})\, d\mathbf{r}.$$

This can be bounded by

$$\langle \tilde{g}^2 \rangle \leq \left(\frac{1-\exp[-\mu b])}{\mu} 4\pi \right)^2 S_1^2 \int f(\mathbf{r})\, d\mathbf{r}. \tag{4.25}$$

Since $\int f(\mathbf{r})\, d\mathbf{r} = 1$, the variance exists.

The actual Monte Carlo algorithm will comprise sampling an $\mathbf{r}$ from $f(\mathbf{r})$; if $\mathbf{r}+\mathbf{X}_0$ is outside the volume V the contribution of $S(\mathbf{r}+\mathbf{X}_0)$ will be zero. The value of the variance obtained in the calculation will be adversely affected by many zeros occurring in the estimate for $\langle S \rangle$. If a pdf defined within a sphere containing V proves to be unsuitable in the calculation because of frequent sampling of $\mathbf{r}$ outside V, it may be necessary to design a rejection technique for $f(\mathbf{r})$ that is nonzero only within V or to sample $\mathbf{r}$ from (4.22) over a smaller sphere and use a different distribution in the rest of V.

Importance sampling is a very useful technique for reducing variance and for handling singular integrals. It will be used repeatedly in the applications of Monte Carlo methods to be discussed. A shortcoming of importance sampling is that it is difficult to apply if the integrand changes sign. In such cases, it will normally be advantageous to use correlated sampling methods.

Suppose we need to evaluate the integrals

$$G_n = \int g_n(x) f(x)\, dx, \tag{4.26}$$

where the $g_n(x)$, $n = 1, 2, \ldots, M$, are a sequence of functions. The most direct method would be to sample an x_i from $f(x)$ and evaluate all the functions for the particular x_i and repeat the process as many times as necessary; that is, sample $x_1, \ldots, x_N$ and form

$$G_{nN} = \frac{1}{N} \sum_{i=1}^{N} g_n(x_i).$$

It is also possible to introduce importance sampling into the integral,

$$G_n = \int \frac{g_n(x) f(x)}{\tilde{f}(x)} \tilde{f}(x)\, dx,$$

sample an x_i from $\tilde{f}(x)$ and form

$$G_{nN} = \frac{1}{N} \sum_{i=1}^{N} \frac{g_n(x_i) f(x_i)}{\tilde{f}(x_i)}.$$

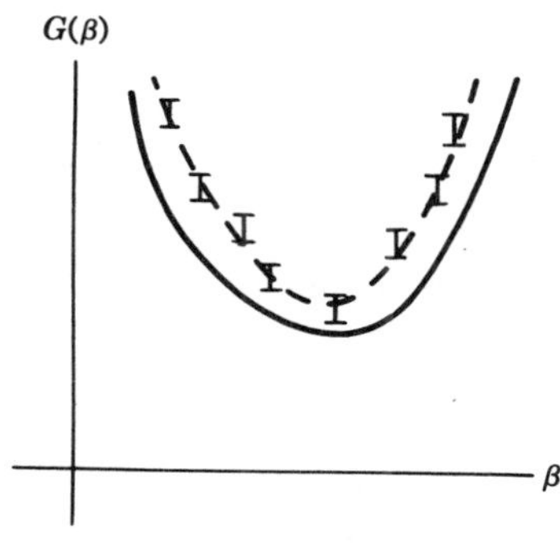

Figure 4.5. Using positive correlation to locate the minimum of an integral $G(\beta)$.

By using the same x_i for all the $g_n(x)$ saves computer time in sampling $f(x)$ or $\tilde{f}(x)$. In many cases the $g_n(x)$ will not differ much; for example, a parameter in the analytical expression for $g_n(x)$ may be changed to observe systematic variations. Positive correlation is introduced into the estimates for G_n by using the same x_i and random fluctuations are eliminated. If we were searching for the parameter β that minimizes the integral $G(\beta)$, the positive correlation makes it easier to locate the minimum β (Figure 4.5).

4.2. THE USE OF EXPECTED VALUES TO REDUCE VARIANCE

The following discussion describes the use of expected values in finite quadrature but its application is much more wide ranging.

Suppose we wish to evaluate the integral

$$G = \int g(X, Y) f(X, Y)\, dX\, dY, \tag{4.27}$$

where X and Y may be many-dimensional vectors (though Y is usually one dimensional). The marginal distribution for X is defined by [Eq. (2.27)]

$$m(X) = \int f(X, Y)\, dY \tag{4.28}$$

and we can define another quantity $h(X)$ as

$$h(X) = \frac{1}{m(X)} \int g(X, Y) f(X, Y) \, dY. \tag{4.29}$$

We assume that the integrals $m(X)$ and $h(X)$ can be evaluated by means other than Monte Carlo. The integral in Eq. (4.27) can be rewritten

$$G = \int m(X) h(X) \, dX. \tag{4.30}$$

We assume that the order of integration is immaterial. The difference in variance between Eqs. (4.27) and (4.30) is (cf. Section 2.5)

$$\begin{aligned} \text{var}\{g\} - \text{var}\{h\} &= E(g^2) - E((E(g \mid X))^2) \\ &= E(E(g^2 \mid X)) - E((E(g \mid X))^2) \\ &= E(E(g^2 \mid X) - (E(g \mid X))^2) = E(\text{var}\{g \mid X\}) \geq 0 \end{aligned} \tag{4.31}$$

and we have shown the general theorem that

$$\text{var}\{g(X, Y)\} - \text{var}\{h(X)\} \geq 0. \tag{4.32}$$

In other words, the variance of a Monte Carlo calculation may be reduced by doing part of the integration analytically as in Eq. (4.29).

As an example, consider the integral

$$\frac{\pi}{4} = \int_0^1 \int_0^1 g(x, y) \, dx \, dy,$$

where

$$g(x, y) = \begin{cases} 1, & x^2 + y^2 \leq 1 \\ 0, & x^2 + y^2 > 1. \end{cases}$$

If we choose x and y uniformly within the unit square and sum $g(x, y)$, the variance associated with the Monte Carlo estimator is

$$\text{var}\{g\} = \frac{\pi}{4} - \left(\frac{\pi}{4}\right)^2 = 0.168.$$

Instead, we can evaluate the marginal distribution for x,

$$m(x) = \int_0^1 dy = 1$$

and define the quantity $h(x)$

$$h(x) = \frac{1}{m(x)} \int g(x, y) f(x, y)\, dy = \int_0^{\sqrt{1-x^2}} dy = \sqrt{1 - x^2}.$$

The integral is rewritten

$$G = \int_0^1 \sqrt{1 - x^2}\, dx.$$

A Monte Carlo algorithm would be to sample x uniformly on (0, 1) and sum the $h(x)$. The variance associated with this is

$$\mathrm{var}\{h\} = \int_0^1 (1 - x^2)\, dx - \left(\frac{\pi}{4}\right)^2 = 0.050,$$

which is indeed a reduction over the previous value. The improvement in the variance will be lost, however, if the second method takes much longer to compute (here, the time for calculation of the square root). The computational time for fixed error can be decreased by using importance sampling to improve the sampling of $\sqrt{1 - x^2}$; the calculation would then be a combination of expected values and importance sampling. In order to use the expected value of the random variable in the Monte Carlo calculation, it must be possible to evaluate $m(x)$ analytically.

Expected values can also be advantageously applied to the Metropolis et al. method. As described previously in Section 3.7, the $\mathrm{M(RT)}^2$ technique generates a random walk in which a possible next step X'_{n+1} is sampled from the transition density $T(X'_{n+1} \mid X_n)$. The probability of accepting the new step is given by

$$A(X'_{n+1} \mid X_n) = \min\left[1, \frac{f(X'_{n+1}) T(X_n \mid X'_{n+1})}{f(X_n) T(X'_{n+1} \mid X_n)}\right]$$

and $X_{n+1} = X'_{n+1}$ with probability $A(X'_{n+1} \mid X_n)$ and $X_{n+1} = X_n$ with probability $1 - A(X'_{n+1} \mid X_n)$. The method guarantees that the asymptotic distribution of the X_i will be $f(X)$. To estimate a quantity G,

$$G = \int g(X) f(X)\, dX,$$

$$G \cong \frac{1}{N} \sum g(X_i),$$

the $g(X_i)$ are summed only after the asymptotic distribution, $f(X)$, has been attained in the random walk. Suppose that we are at step X_n in the random walk and have sampled a possible next step X'. Conditional on X_n, the expectation value for $g(X_{n+1})$ is

$$E(g(X_{n+1})) = \int_0^1 \int_0^1 g_1(X', Y) T(X' \mid X_n)\, dX'\, dY,$$

where

$$g_1(X', Y) = \begin{cases} g(X') & \text{if} \quad Y \leq A(X' \mid X_n) \\ g(X_n) & \text{if} \quad Y > A(X' \mid X_n). \end{cases}$$

The variable Y is a uniform random variable and the Monte Carlo calculation may be improved by using its expected value in the algorithm. The marginal distribution for X' is

$$m(X') = \int_0^1 T(X' \mid X_n)\, dY = T(X' \mid X_n)$$

and $h(X')$ becomes from Eq. (4.29)

$$\begin{aligned} h(X') &= \frac{1}{T(X' \mid X_n)} \int_0^1 g_1(X', Y) T(X' \mid X_n)\, dY \\ &= \frac{1}{T(X' \mid X_n)} \left[\int_0^A g(X') T(X' \mid X_n)\, dY + \int_A^1 g(X_n) T(X' \mid X_n)\, dY \right] \\ &= g(X') A(X' \mid X_n) + g(X_n)[1 - A(X' \mid X_n)]. \end{aligned}$$

That is, the scores at X' and X_n are weighted with the probabilities that X' or X_n will be the next point.

In some applications $A(X' \mid X_n)$ is either 0 or 1 (e.g., in sampling classical many-body systems with hard sphere forces), so there is no advantage in using expected values. There are many applications in which $A(X' \mid X_n)$ varies smoothly between 0 and 1 and the method can be very useful. Usually, the most computer time in a $M(RT)^2$ algorithm is spent calculating $f(X')$ and $f(X_n)$ for the acceptance probability. The calculation becomes more efficient if $g(X')$ and $g(X_n)$ are determined simultaneously with $f(X')$ and $f(X_n)$. In conventional $M(RT)^2$ applications, one of the quantities, $g(X')$ or $g(X_n)$, is then thrown away, whereas if expected values are used, both values are needed. In the earlier example of disks in a box, $f(X)$ is determined by checking whether any disks overlap or are outside the box. Since all separations of pairs of disks must be calculated for $f(X)$, it is most efficient to keep track of the occurrences of a particular separation [which is $g(X)$] at the same time. The expected values method will be useful if the disks are not subject to perfect repulsion and especially if $g(X)$ is sensitive to the effects of small distances.

4.3. CORRELATION METHODS FOR VARIANCE REDUCTION

Correlation methods serve to reduce the variance by the use of correlated points in the sampling rather than sampling all points independently. In a technique called *control variates*,[1] the integral of interest,

$$G = \int g(x) f(x)\, dx$$

is written

$$G = \int (g(x) - h(x)) f(x)\, dx + \int h(x) f(x)\, dx \tag{4.33}$$

where $\int h(x) f(x)\, dx$ is known analytically. The estimator for G becomes

$$G \cong \int h(x) f(x)\, dx + \frac{1}{N} \sum_{i=1}^{N} [g(x_i) - h(x_i)], \tag{4.34}$$

with $g(x)$ and $h(x)$ evaluated at the same points (x_i). The technique is advantageous when

$$\operatorname{var}\{(g(x) - h(x))\}_f \ll \operatorname{var}\{g(x)\}_f, \tag{4.35}$$

and this occurs when $h(x)$ is very similar to $g(x)$. If $\int g(x)f(x)\,dx$ closely resembles a known integral, then the method will probably be useful. In particular, if $|g(x)-h(x)|$ is approximately constant for different values of $h(x)$, then correlated sampling will be more efficient than importance sampling.[2] Conversely, if $|g(x)-h(x)|$ is approximately proportional to $|h(x)|$, then importance sampling would be the appropriate method to use.

Consider the integral

$$\int_0^1 e^x\,dx;$$

a straightforward Monte Carlo calculation gives a variance equal to 0.242. A possible $h(x)$ derives from the first two terms in the Taylor series of e^x,

$$\int_0^1 e^x\,dx = \int_0^1 (e^x - (1+x))\,dx + \tfrac{3}{2},$$

where $\int_0^1 (1+x)\,dx = \frac{3}{2}$. The random variable x may be chosen uniformly on (0, 1) and the associated variance is

$$\mathrm{var}\{e^x - (1+x)\} = 0.0437,$$

which is a substantial reduction from the variance quoted above. By using $\frac{2}{3}(1+x)$ as an importance function, however, the variance is decreased to 0.0269.

As a better example, consider the integral

$$\int_0^1 (e^x - (1+\beta x))\,dx,$$

where we shall minimize the variance with respect to β. We want the value of β for which

$$\int_0^1 (e^x - 1 - \beta x)^2\,dx - \left(\int_0^1 (e^x - 1 - \beta x)\,dx\right)^2$$

is smallest. Experimentally it can be determined that $\beta = 1.69$ gives a variance of 0.0039. The best fit to the exponential in the sense given here is not far from the best fit for a straight-line importance function. In the latter case, $\beta = 1.81$ is optimal and the variance is 0.0040, rather close to the best value found here.

In practice, the experimental determination of an optimal parameter or parameters for a control variate treatment of a many-dimensional integral cannot be carried out with great precision, since determination of the variance to be minimized must itself be determined by Monte Carlo.

Control variates are used extensively in simulations,[3] particularly in the study of queues and queuing networks.

4.4. ANTITHETIC VARIATES

The method of antithetic variates[4] exploits the decrease in variance that occurs when random variables are negatively correlated. When variables are negatively correlated, if the first point gives a value of the integrand that is larger than average, the next point will be likely to give a value that is smaller than average, and the average of the two values will be closer to the actual mean. In theory the method sounds promising, but in practice it has not been very successful as a general variance reduction method in many dimensions.[5] It can often be used to advantage for a few variables of the many.[6]

Suppose that

$$G = \int_0^1 g(x)\, dx, \tag{4.36}$$

where $g(x)$ is linear. Then G may be written exactly as

$$G = \int_0^1 \tfrac{1}{2}[g(x) + g(1-x)]\, dx. \tag{4.37}$$

G may be evaluated through Monte Carlo by using Eq. (4.37) and picking an x uniformly on $(0, 1)$. The value of $g(x)$ and $g(1-x)$ is determined and the estimate for G formed,

$$G_N = \frac{1}{N} \sum_{i=1}^{N} \tfrac{1}{2}[g(x_i) + g(1-x_i)], \tag{4.38}$$

which will give exactly G with zero variance for linear g (Figure 4.6). For nearly linear functions, this method will substantially reduce the variance.

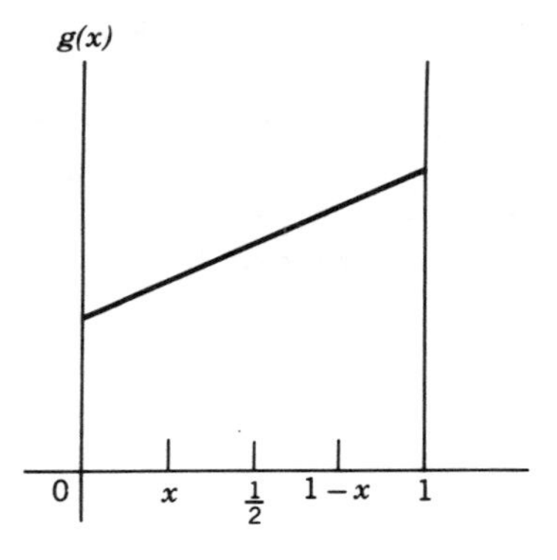

Figure 4.6. A linear function $g(x)$ on $(0, 1)$.

Consider the integral discussed earlier,

$$G = \int_0^1 e^x \, dx;$$

the variance associated with a straightforward Monte Carlo evaluation is 0.242. If we pick $x_1, x_2, \ldots, x_N$ uniformly and at random on $(0, 1)$ and form the estimator in Eq. (4.38)

$$G_N = \frac{1}{N} \sum_{n=1}^{N} \tfrac{1}{2}[g(x_n) + g(1 - x_n)],$$

the variance associated with this calculation is 0.0039, which is a dramatic reduction in variance.

As another example take

$$G = \int_0^\infty e^{-x} g(x) \, dx.$$

We shall use e^{-x} as the sampling function. A random number distributed exponentially is given by

$$x = -\log \xi.$$

A correlated random variable is $x' = -\log(1 - \xi)$. If x is close to the origin, x' will be far from the origin. The values for x and x' can be substituted in the expression for the estimator in Eq. (4.38), which becomes

$$G = \frac{1}{N} \sum_{i=1}^{N} \tfrac{1}{2}\{g(x_i) + g(x_i')\}.$$

This will be an improvement over the use of a single estimate if g is monotone. Two generalizations of Eq. (4.38) may be introduced, for any $0 < \alpha < 1$:

$$g_{+\alpha} = \{\alpha g(\alpha x) + (1-\alpha) g(\alpha + (1-\alpha)x)\}$$

and

$$g_{-\alpha} = \{\alpha g(\alpha x) + (1-\alpha) g(1 - (1-\alpha)x)\}; \tag{4.39}$$

these functions are linear and have the correct expectation. A value for α may be established by solving

$$\begin{aligned} g_{-\alpha}(0) &= g_{-\alpha}(1), \\ g(\alpha) &= (1-\alpha) g(1) + \alpha g(0). \end{aligned} \tag{4.40}$$

Correlation techniques need not be used alone, but may be combined with other methods of reducing the variance. For example, both importance sampling and antithetic variables may be used simultaneously to improve a calculation. Thus the result expressed in Eq. (4.37) may be improved by importance sampling as

$$G = \frac{1}{2} \int_0^1 \left[\frac{g(x) + g(1-x)}{\tilde{f}(x)} \right] \tilde{f}(x)\, dx. \tag{4.41}$$

Consider again $\int_0^1 e^x\, dx$. The antithetic estimator $\frac{1}{2}[e^x + e^{1-x}]$ is symmetric about $x = \frac{1}{2}$, so an approximate $\tilde{f}$ for this problem is

$$\tilde{f}(x) = \tfrac{24}{25}[1 + \tfrac{1}{2}(x - \tfrac{1}{2})^2],$$

chosen to agree with three terms of the power series at $x = \frac{1}{2}$. The variance of the estimator of Eq. (4.41) with this sampling function is 0.0000012, compared with 0.0039 for the estimator of Eq. (4.37) and 0.242 for the straightforward evaluation. Thus, use of antithetic variates reduces the variance by two orders of magnitude, and this simple unoptimized importance sampling by another three!

4.5. STRATIFICATION METHODS

In stratification methods the domain of the independent variable is broken up into subintervals such that a random variable is sampled from every subinterval. For the one-dimensional integral

$$G = \int_0^1 g(x)\, dx, \tag{4.42}$$

the simplest stratification is to divide (0, 1) into M equal intervals. An x is then chosen in each interval in succession,

$$x_m = \frac{l - \xi_m}{M},$$

where $l = (m-1)(\text{mod } M) + 1$ and $m = 1, 2, \ldots, NM$. The l's will cycle through the integer values 1 through M, and the x_m will be drawn at random in the lth interval. An estimator for G is given by

$$G_{NM} = \frac{1}{NM} \sum g(x_m). \tag{4.43}$$

It is best, however, not to use (4.43) as written, since estimation of the variance is difficult. Rewriting G_{NM} as

$$G_{NM} = \frac{1}{N} \sum_n \left(\frac{1}{M} \sum_m g(x_m) \right),$$

gives terms $(1/M) \sum g(x_m)$ that are statistically independent. The variance is straightforward to calculate from these. In correlation methods it is very important to group answers such that the individual groups are independent in order to calculate the variance. For example, in antithetic variates we group two (or more) answers as an independent estimate in the determination of the variance.

An alternative method for stratifying the sampling of an x for the integral in Eq. (4.42) would be to use the same random number in each subinterval per cycle of l. That is

$$x_m = \frac{l - \xi_k}{M},$$

where

$$l = (m-1)(\text{mod } M) + 1,$$

$$k = \left[\frac{m-1}{M}\right] + 1,$$

and m is defined as before. In one dimension, this is analogous to trapezoidal integration, but the random choice of x on a subinterval removes a bias.

There is no particular reason for choosing the subintervals to be of equal size. An estimator g_k for the integral in (4.42) can be defined that allows varying subinterval size,

$$g_k = \sum_{j=1}^{k} \sum_{i=1}^{n_j} (\alpha_j - \alpha_{j-1}) \frac{1}{n_j} g[\alpha_{j-1} + (\alpha_j - \alpha_{j-1})\xi_{ij}]. \tag{4.44}$$

The notation ξ_{ij} indicates that a new random number is chosen for every combination of (i, j); the random number is then mapped onto the interval (α_{j-1}, α_j) in which n_j samples are to be used. The sampling algorithm maps (i, j) onto the interval (α_{j-1}, α_j). A particular subinterval (α_{j-1}, α_j) is sampled n_j times, and the mean of the g's evaluated within (α_{j-1}, α_j) is multiplied by the size of the interval. This is a generalization of the trapezoidal rule for integration since the size of the interval changes. The variance on a subinterval is proportional to the number of samples taken within the subinterval

$$n_j^2 \propto \left\{(\alpha_j - \alpha_{j-1}) \int_{\alpha_{j-1}}^{\alpha_j} g^2(x)\,dx - \left[\int_{\alpha_{j-1}}^{\alpha_j} g(x)\,dx\right]^2\right\}.$$

It is possible to minimize the variance with respect to n_j to determine the optimal value of n_j.

It is not really necessary to use variable interval size stratified sampling since an importance sampling transformation will achieve the same goal. The motivation to vary the subinterval size is to have large intervals where the function is nearly constant and fine intervals where the function is large or rapidly changing. But importance sampling leads to the same result by sampling a function that is large where f is large. A useful approach is to importance-sample an integral and then use stratified sampling on the transformed integral.

A straightforward generalization of stratified sampling to many dimensions is not efficient, since it leads to the necessity of sampling in

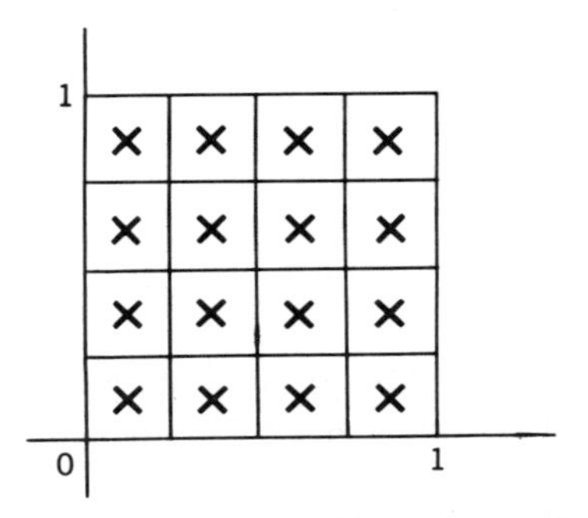

Figure 4.7. Dividing the unit square into 16 subintervals.

many cells. Rather than attempt to stratify all the dimensions, it is better to identify which variables (if any) carry the majority of the variation of the integrand and stratify these. Significant reduction in the variance can sometimes be achieved by stratifying a single dimension in a many-dimensional integral.

Suppose we wish to sample random variables from the unit square in two dimensions. The generalization of simple stratification into four intervals in one dimension would be to divide the square into 16 subintervals and sample a point in each box (Figure 4.7). An alternative of this has been suggested by Steinberg[7] in which points are sampled in only four boxes, one from each row. The points would be chosen in boxes $(1, n_1)$, $(2, n_2)$, $(3, n_3)$, $(4, n_4)$, where the sequence n_1, n_2, n_3, n_4 is a random permutation of $1, 2, 3, 4$. A particular permutation of the sequence is shown in Figure 4.8. The sampled boxes will stay apart in this scheme, and no accidental sampling of two random variables in the same subinterval will occur. We see that it is always perfectly stratified with respect to x and y separately, which is better than arbitrarily choosing a single direction and then stratifying. In two dimensions Steinberg's

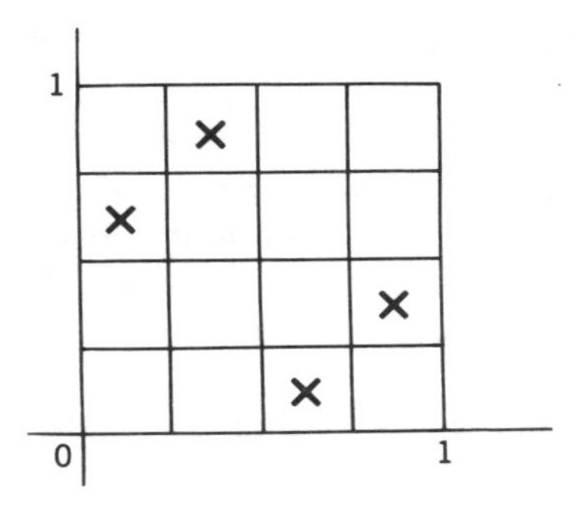

Figure 4.8. Sampling one subinterval from each row in the unit square by a random permutation.

scheme is not usually an improvement, but its generalization to many dimensions permits the retention of much of the advantage of stratification with reasonably small groups of correlated variables.

Specifically, suppose we wish to stratify k samples in a d-dimensional unit hypercube. Let each of the sides be divided into k segments (for a total of k^d small hypercubes). Form $d-1$ independent permutations of the integers $1, 2, \ldots, k$. Let the ith members of the jth permutation be $n(i, j)$. Assemble k d-tuples as

$$[1, n(1, 1), n(1, 2), n(1, 3), \ldots, n(1, d-1)],$$

$$[2, n(2, 1), n(2, 2), n(2, 3), \ldots, n(2, d-1)], \qquad \ldots,$$

$$[k, n(k, 1), n(k, 2), \ldots, n(k, d-1)],$$

and place a sample at a point in each of the small hypercubes so defined. As before, the points can but need not be independently sampled within each hypercube.

4.6. GENERAL-PURPOSE MONTE CARLO INTEGRATION CODE

Some research has gone into the writing of general-purpose multidimensional Monte Carlo integration codes. The codes attempt to deal with badly behaved functions in a reasonable way. An example is a code DIVONNE2.[8] The code consists of two separate programs; the first performs a recursive multidimensional stratification of the function space. Then the second program carries out a Monte Carlo calculation based on the stratification done by the first. The goal of the partitioning is to produce subvolumes in which the range of function values is as small as possible.

4.7. COMPARISON OF MONTE CARLO INTEGRATION AND NUMERICAL QUADRATURE

In the introduction to this chapter, the relative efficiencies of a Monte Carlo calculation as compared with standard numerical quadrature were discussed. A recent review[9] of this subject has developed criteria to use to judge whether a Monte Carlo integration is appropriate. Several points are worth emphasizing. For fixed expenditure of computer time, numerical quadrature is always the wiser choice in one-dimensional integration. In higher dimensions, however, the convergence of Monte Carlo integrations is independent of dimensionality, which is not true for numerical quadrature methods. Thus there is always some dimension value d above which Monte Carlo integrations converge faster than any fixed quadrature rule. Also, the shape of the multidimensional region may be quite complicated and may make applications of a standard quadrature routine difficult. Monte Carlo methods are able to deal with essentially any finite region. Finally, the accuracy of a Monte Carlo estimate of an integral can always be improved by simply including some more points. In many dimensions, a fixed-point numerical quadrature rule can be improved only by going to a higher-order rule, and the previous estimate is thrown out! Error estimates for multidimensional numerical quadrature are difficult to evaluate, whereas a Monte Carlo calculation always delivers an associated error.

REFERENCES

1. M. Kahn and A. W. Marshall, Methods of reducing sample size in Monte Carlo computations, *Oper. Res.*, **1**, 263, 1953.
2. J. H. Halton, On the relative merits of correlated and importance sampling for Monte Carlo integration, *Proc. Camb. Phil. Soc.*, **61**, 497, 1965.
3. S. S. Lavenburg and P. D. Welch, A perspective on the use of control variates to increase the efficiency of Monte Carlo simulations, Research Report RC8161, IBM Corp., Yorktown Heights, New York, 1980.
4. J. M. Hammersley and K. W. Morton, A new Monte Carlo technique antithetic variates, *Proc. Camb. Phil. Soc.*, **52**, 449, 1956.
5. J. H. Halton and D. C. Handscomb, A method for increasing the efficiency of Monte Carlo integration, *ACM J.*, **4**, 329, 1957.
6. J. H. Halton, Generalized Antithetic Transformations for Monte Carlo Sampling, Computer Sciences Technical Report 408, University of Wisconsin, Madison, 1980.
7. H. A. Steinberg, MAGI Corp., White Plains, New York, private communication.
8. J. H. Friedman and M. H. Wright, A nested partitioning procedure for numerical multiple integration, *ACM Trans. Math. Software*, **7**, 76, 1981.
9. F. James, Monte Carlo Theory and Practice, *Rep. Prog. Phys.*, **43**, 1145, 1980.

GENERAL REFERENCES

J. H. Halton, A retrospective and prospective survey of Monte Carlo methods, *Soc. Indust. Appl. Math. Rev.*, **12**, 1, 1970.

F. James, Monte Carlo Theory and Practice, *Rep. Prog. Phys.*, **43**, 1145, 1980.

5 STATISTICAL PHYSICS

In the next few chapters we give introductions to Monte Carlo methods as they are used in specific applications. This is, of course, important in itself, but many technical problems that arise in the applications are difficult to motivate out of context. Thus in this chapter we discuss applications in physics that require the evaluation of (usually very) many-dimensional integrals. The $M(RT)^2$ or Metropolis method, introduced in Section 3.7 is essential for practical treatment, so here we shall illustrate some technical points and discuss some of the scientific ideas which underlie the numerical simulations of physical systems.

5.1. CLASSICAL SYSTEMS

Suppose that $\mathbf{R}$ is a many-dimensional vector whose associated probability distribution function is $f(\mathbf{R})$. The $M(RT)^2$ method constructs a random walk such that the points $\mathbf{R}_i$ will eventually be drawn from $f(\mathbf{R})$. The random walk is initiated by choosing $\mathbf{R}_1$ from a pdf $\phi_1(\mathbf{R}_1)$. A possible next point is sampled using the transition density $T(\mathbf{R}_2' \mid \mathbf{R}_1)$ and the decision whether to accept $\mathbf{R}_2'$ is based on the quantity q, where

$$q = \frac{f(\mathbf{R}_2')\,T(\mathbf{R}_1 \mid \mathbf{R}_2')}{f(\mathbf{R}_1)\,T(\mathbf{R}_2' \mid \mathbf{R}_1)}. \tag{5.1}$$

If $q > 1$, then $\mathbf{R}_2' = \mathbf{R}_2$. If $q < 1$, $\mathbf{R}_2'$ is accepted with probability q; otherwise $\mathbf{R}_2 = \mathbf{R}_1$. This process is repeated many times; each time a next $\mathbf{R}_i'$ is chosen from the transition density and accepted depending on the value of q. The eventual distribution of $\mathbf{R}$'s is guaranteed to be $f(\mathbf{R})/\int f(\mathbf{R})\,d\mathbf{R}$.

Problems in statistical physics usually deal with the properties and characteristics of large ensembles of particles. For example, a classical situation[1] is N particles confined in a box. The configuration of the system may be represented by the multidimensional vector $\mathbf{R}$, which contains the $3N$ coordinates $\mathbf{r}_i = (x_i, y_i, z_i)$ of the centers of mass of the particles in the box.

The total energy of the system is the sum of the kinetic energies of all the particles and the potential energy

$$E = \sum_{i=1}^{N} \frac{p_i^2}{2m_i} + \sum_{i<j\leq N} \Phi(|\mathbf{r}_i - \mathbf{r}_j|). \tag{5.2}$$

Here p_i is the momentum and m_i is the mass of particle i. We have assumed that the total potential energy is a sum of the potential energies between a pair of particles. Typical pair potentials used in calculations of this sort are the hard sphere potential (shown in Figure 5.1):

$$\Phi_{hs}(r) = \begin{cases} \infty, & r < r_0 \\ 0, & r \geq r_0, \end{cases}$$

and the Lennard-Jones potential (shown in Figure 5.2)

$$\Phi_{LJ}(r) = 4\epsilon\left[\left(\frac{\sigma}{r}\right)^{12} - \left(\frac{\sigma}{r}\right)^{6}\right],$$

where r is the separation between two particles. The Lennard-Jones potential is a reasonably good representation of the pair potential[2] in simple liquids such as helium, argon, and neon. The distribution of the

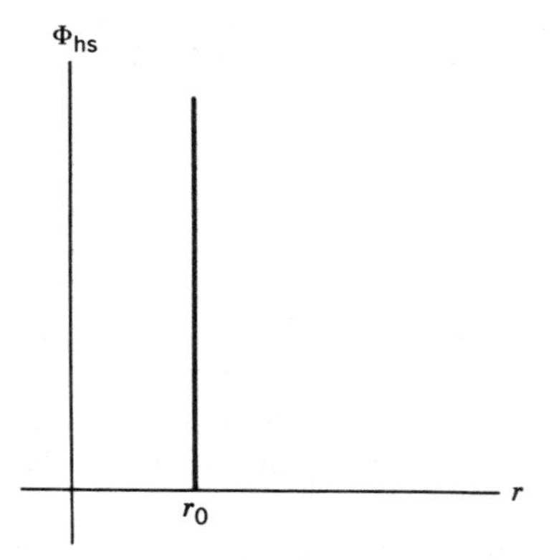

Figure 5.1. The hard sphere potential energy function.

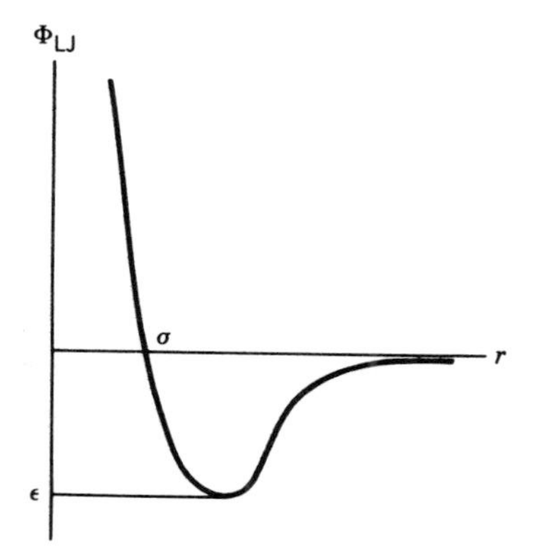

Figure 5.2. The Lennard-Jones potential energy function.

particles in the box is a function of the particle positions and velocities:

$$f(v, \mathbf{R}) = \exp[-E(v, \mathbf{R})/k_b T] / \int \exp[-E(v, \mathbf{R})/k_b T]\, d\mathbf{R}\, dv, \quad (5.3)$$

where k_b is Boltzmann's constant, T is the temperature, and $E(v, \mathbf{R})$ is the energy of the system. At low temperatures the particles will be close to some minimum energy configuration (solid) and at high temperatures $(T \to \infty)$, $f(v, \mathbf{R}) = \text{const}$ and the particles will be uniformly distributed as a gas. The distribution function in Eq. (5.3) can be rewritten in terms of relative coordinates (interparticle separations, $r_{ij} = |\mathbf{r}_i - \mathbf{r}_j|$) and the velocities can be integrated out to yield the Boltzmann distribution function

$$f(\mathbf{R}) = \frac{\exp[-\sum \Phi(r_{ij})/k_b T]}{\int \exp[-\sum \Phi(r_{ij})/k_b T]\, d\mathbf{R}}. \quad (5.4)$$

We are now ready to determine the average energy of the particles, but we cannot do it analytically. A possible procedure is to sample $f(\mathbf{R})$ through the $M(RT)^2$ method and determine the energy by Monte Carlo, that is, to estimate the average of $\sum_{i<j} \Phi(r_{ij})$.

5.1.1. The Hard Sphere Liquid

Suppose we have a hard sphere liquid in a three dimensional box (Figure 5.3), where each hard sphere of diameter a moves in a cube of dimension

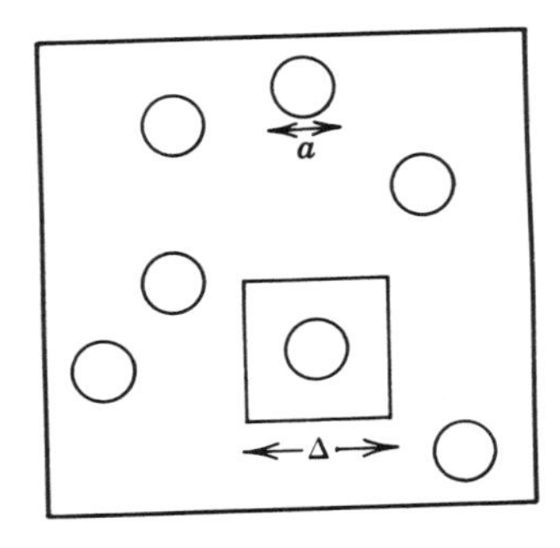

Figure 5.3. A two dimensional slice through a hard sphere liquid.

Δ. As transition density in M(RT)[2] we shall choose

$$T(\mathbf{R}'_n|\mathbf{R}_{n-1}) = \begin{cases} \dfrac{1}{\Delta^3} & \text{for } |x'_i - x_i|,\ |y'_i - y_i|,\ |z'_i - z_i| < \Delta/2,\ r'_j = r_j,\ j \neq i, \\ 0 & \text{otherwise,} \end{cases}$$

where i is chosen at random from 1 to N. The constant transition density drops out of the quotient for q (Eq. (5.1)). If a system can go from $\mathbf{R}_1$ to $\mathbf{R}'_2$, the reverse move is equally possible. Equation (5.1) becomes

$$q = \frac{\exp[-\sum \Phi(r'_{ij})/k_b T]}{\exp[-\sum \Phi(r_{ij})/k_b T]} = \exp\left[\frac{-\Delta U}{k_b T}\right],$$

where $\Delta U = \sum \Phi(r'_{ij}) - \sum \Phi(r_{ij})$. A move to a region of lower potential energy is automatically accepted; otherwise, the move is accepted with probability q.

In most cases we are trying to model an infinite liquid, and the box size can have a dramatic effect on the final answer.[3] This may be partly overcome by using periodic boundary conditions on the sides of the box. Periodic boundary conditions are illustrated in Figure 5.4. When a

Figure 5.4. A particle moving beyond the edge of the box appears at its image position within the box.

particle moves past the box on one side, it appears at the same position (less the width of the box) on the "opposite" side, and the potential energy is computed using some or all of the image positions obtained by translating through the dimensions of the box.

One property of the liquid that can be calculated through the Monte Carlo simulation is the radial distribution function. The radial distribution function $g(r)$ is the probability that two particles are a distance r apart,

$$g(r) = \left(1 - \frac{1}{N}\right)\left\langle \sum_{i<j} \delta(r - |\mathbf{r}_i - \mathbf{r}_j|)\right\rangle. \tag{5.5}$$

In the Monte Carlo simulation it is determined by simply recording in narrow bins the frequency of occurrence of different particle separations. For hard spheres, $g(r)$ has the behavior that the particles tend to clump together such that many have an interparticle separation a little larger than the hard sphere diameter. This is shown in Figure 5.5. The radial distribution function can be used to determine the potential energy of the system when the potential is Φ (not a hard sphere)

$$U = \frac{\rho}{2} \int g(r)\Phi(r)\, d^3r, \tag{5.6}$$

where ρ is the density of the system. Other properties of liquids that may be calculated during the random walk are the structure function $S(k)$ [the Fourier transform of $g(r)$], the specific heat, and the density. However, it is not at all easy to determine directly A, the free energy of the system, by this kind of simulation:

$$Q = \exp\left[-\frac{A}{k_b T}\right] = \frac{\int \exp[-\sum \Phi(r_{ij})/k_b T]\, d\mathbf{R}}{V^N}, \tag{5.7}$$

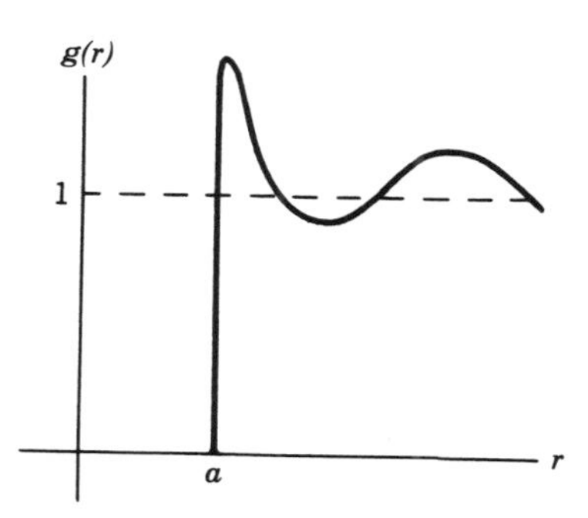

Figure 5.5. The radial distribution function for hard spheres.

where V is the volume and Q is the canonical configuration integral. The free energy can be found using either a series of simulations at different temperatures[4] or a correlational Monte Carlo technique[5] that relates A for one system to the known A for a different one.

5.1.2. Molecular Dynamics

An alternative method used to study liquid systems is molecular dynamics.[6] In molecular dynamics the motion of the particles is simulated by solving numerically Newton's equations for the system:

$$\mathbf{F} = m\frac{d\mathbf{v}_i}{dt} = -\nabla_i \sum_{j<k} \Phi(r_{jk}), \tag{5.8}$$

where $\mathbf{v}_i$ is a particle velocity. The calculation is started with the particles distributed on a lattice and initial velocities are randomly selected from the Boltzmann distribution

$$f(v) = \exp[-\tfrac{1}{2}mv^2/k_b T]. \tag{5.9}$$

The equations of motion are integrated in time to determine the paths the particles follow, and from this, time-dependent phenomena may be studied. For static correlation properties, however, Monte Carlo calculations are better since they generally converge faster to the probability distribution function $f(\mathbf{R})$. Also, the Monte Carlo calculations do not have to sample all of phase space as molecular dynamics calculations will, since the integrations over momenta are already done.

5.1.3. The Ising Model

A very well-known model used in statistical physics is the Ising model[7] of interacting spins on a lattice. The model can be used to simulate the properties of a magnet. It is assumed that the spins can either be up or down with the value +1 associated with up and the value −1 associated with down:

$$\text{spin} = \sigma_i = \pm 1, \qquad 1 \le i \le N,$$

where N is the number of particles. The total energy of the system is assumed to be

$$E = \sum_{i<j} J_{ij}\sigma_i\sigma_j + h\sum_{i=1}^{N} \sigma'_i, \tag{5.10}$$

where the second term in Eq. (5.10) derives from an external magnetic field of strength h. The quantity J_{ij} represents the interaction between nearest neighbors; the most commonly used assumption is

$$J_{ij} = \begin{cases} 0 & \text{if } i \text{ and } j \text{ are not nearest neighbors} \\ \text{const} & \text{otherwise.} \end{cases}$$

The Ising model can be solved analytically[8] in two dimensions, but not in three. In either case the Ising model exhibits a phase transition as the temperature is reduced from infinity. In three dimensions this transition has interesting features in common with the critical point of a liquid. The Ising model is also used to model an alloy in which "spins" stand for one or another type of atom that may reside at a lattice site. Clusters of such atoms may develop, depending on the temperature. The $M(RT)^2$ method is commonly used, and various correlation functions can be extracted from the random walk.

Monte Carlo techniques have been used to study exhaustively systems obeying spherical potentials and are now being applied to more complicated systems. For example surfaces can be studied by constructing a liquid and then removing the boundaries to form droplets.[9] A liquid like water can be simulated by using more complicated potentials that are angle dependent.[10] Liquid crystals[11] are being studied by changing the particles in a box from disks into rods.

5.2. QUANTUM SIMULATIONS

The simulations described above have all assumed that the particles involved obey classical mechanics. It is also possible to study systems that are essentially quantum-mechanical in nature. The behavior of a quantum-mechanical system is described by the Schrödinger equation,

$$H\Psi(\mathbf{R}) = E_0\Psi(\mathbf{R}), \tag{5.11}$$

where H is the hamiltonian for the system,

$$H = \sum_{i=1}^{N} -\frac{\hbar^2}{2m}\nabla_i^2 + \sum_{i<j}\Phi(r_{ij}), \qquad r_{ij} = |\mathbf{r}_i - \mathbf{r}_j|,$$

$$\nabla_i^2 = \frac{\partial^2}{\partial x_i^2} + \frac{\partial^2}{\partial y_i^2} + \frac{\partial^2}{\partial z_i^2}, \tag{5.12}$$

$\hbar$ is Planck's constant, $\Psi(\mathbf{R})$ is the wave function, and E_0 is the ground-state energy. The quantity ∇_i^2 determines the kinetic energy of particle i, and $\Phi(r_{ij})$ represents the pair potential between particles i and j. At zero temperature, quantum particles are still in motion; this can be contrasted with classical mechanics, where particles are fixed at a minimum potential energy at zero temperature. The Schrödinger equation has been solved analytically only for the hydrogen atom; more complicated systems must be studied numerically.

For few-body and many-body systems, variational methods may be used, and the necessary integrals carried out by Monte Carlo.[12] If Ψ_T is a trial wave function, the variational energy is defined as

$$E_{\text{var}} \equiv \frac{\int \Psi_T(\mathbf{R}) H \Psi_T(\mathbf{R})\, d\mathbf{R}}{\int |\Psi_T(\mathbf{R})|^2\, d\mathbf{R}} \geq E_0. \tag{5.13}$$

That is, E_{var} is an upper bound to the ground-state energy. If $\Psi_T = \Psi(\mathbf{R})$, the actual wave function, then $E_{\text{var}} = E_0$. To use Monte Carlo techniques, the following probability distribution function is very convenient:

$$f(\mathbf{R}) = \frac{|\Psi_T(\mathbf{R})|^2}{\int |\Psi_T(\mathbf{R})|^2\, d\mathbf{R}}, \tag{5.14}$$

and the variational energy [Eq. (5.13)] becomes

$$E_{\text{var}} = \int f(\mathbf{R}) E(\mathbf{R})\, d\mathbf{R} = \langle E(\mathbf{R}) \rangle \geq E_0, \tag{5.15}$$

where $E(\mathbf{R})$ is the *local energy*, defined as

$$E(\mathbf{R}) = \frac{1}{\Psi_T(\mathbf{R})} H \Psi_T(\mathbf{R}). \tag{5.16}$$

To determine an estimate for E_{var}, values for $\mathbf{R}$ are sampled from $f(\mathbf{R})$, and the average of the resulting $E(\mathbf{R})$ is constructed. For systems with more than a few particles, the sampling can be done conveniently only with the $M(RT)^2$ method.

The Monte Carlo variational method has been applied to liquid ^{4}He.[13,14] Liquid ^{4}He is a superfluid at low temperatures and solidifies only at high pressure; its properties must be obtained from a quantum-mechanical calculation. As a trial function, the following functional form is most generally used:

$$\Psi_T(\mathbf{R}) = \prod_{i<j} f(r_{ij}) \tag{5.17}$$

with

$$f(r_{ij}) = \exp[-\tfrac{1}{2}u(r_{ij})], \tag{5.18}$$

where $u(r_{ij})$ is called the pseudopotential since it has a role similar to the potential in the Boltzmann distribution. On substituting in the expression for $f(r_{ij})$ the trial function becomes

$$\Psi_T(\mathbf{R}) = \exp\left[-\frac{1}{2}\sum_{i<j} u(r_{ij})\right]. \tag{5.19}$$

The form of the pseudopotential is arbitrary as long as $u(\infty) = 0$, $u(0) = \infty$, and the first derivative is continuous. The actual form for $u(r_{ij})$ is determined by minimizing E_{var} with respect to u. By applying Green's theorem to Eq. (5.15), E_{var} is written

$$E_{\text{var}} = \frac{\rho}{2}\int d^3r\, g(r)\left\{\Phi(r) + \frac{\hbar^2}{2m}\nabla^2 u(r)\right\}, \tag{5.20}$$

where ρ is the density of the system and $g(r)$ is the radial distribution function [Eq. (5.5)]. The integral in Eq. (5.20) is minimized with respect to $u(r)$.

Another interesting property that can be investigated by the Monte Carlo calculation is the momentum distribution in the liquid, $n(k)$, that is, the probability density that a particle has momentum k. It has been suggested that the reason ^{4}He is a superfluid is that many particles are at rest (the zero momentum state). Classically, the momenta would be distributed according to the Boltzmann distribution [Eq. (5.9)] shown in Figure 5.6. A large fraction of the particles in the zero momentum state would correspond to a delta function at $k = 0$, as indicated in Figure 5.7. Variational Monte Carlo calculations on liquid ^{4}He have indeed observed a substantial fraction of the particles in the zero momentum state. Experimental studies have also derived estimates of this quantity.[15]

Liquid ^{4}He is a boson fluid, which means the wave function is symmetric upon interchange of two particle coordinates. Other interesting particles, for example electrons, are described by an antisymmetric

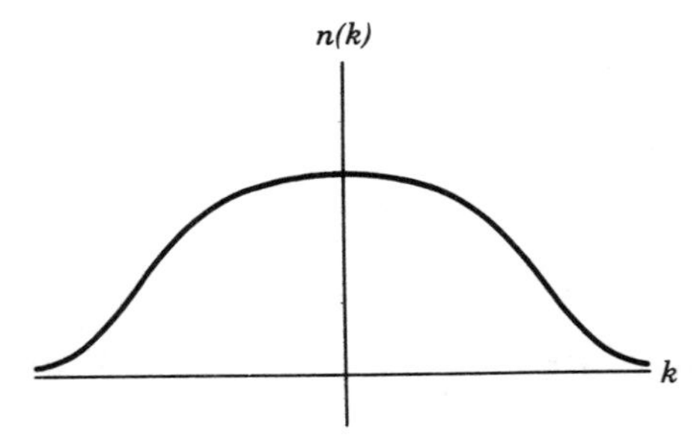

Figure 5.6. Boltzmann distribution of momenta.

wave function and are called *fermions*. The requirement of an antisymmetric wave function makes the description of fermions much more complicated. To treat fermions variationally, the following trial function has been used[12]

$$\Psi_{\mathrm{T}} = \exp\left[-\sum_{i<j} u(r_{ij}) \right] \times \mathrm{Det}[e^{i\mathbf{k}_j \cdot \mathbf{r}_i}], \tag{5.21}$$

where Det[$\cdot\cdot\cdot$] is the determinant of the ideal gas wave function:

$$\mathrm{Det}[e^{i\mathbf{k}_j \cdot \mathbf{r}_i}] = \begin{vmatrix} e^{i\mathbf{k}_1 \cdot \mathbf{r}_1} & e^{i\mathbf{k}_2 \cdot \mathbf{r}_1} & \cdots \\ e^{i\mathbf{k}_1 \cdot \mathbf{r}_2} & e^{i\mathbf{k}_2 \cdot \mathbf{r}_2} & \cdots \\ \vdots & \vdots & \end{vmatrix}.$$

The k_i are momentum vectors in three-dimensional momentum space. When Eq. (5.21) is used in the variational calculations, problems occur

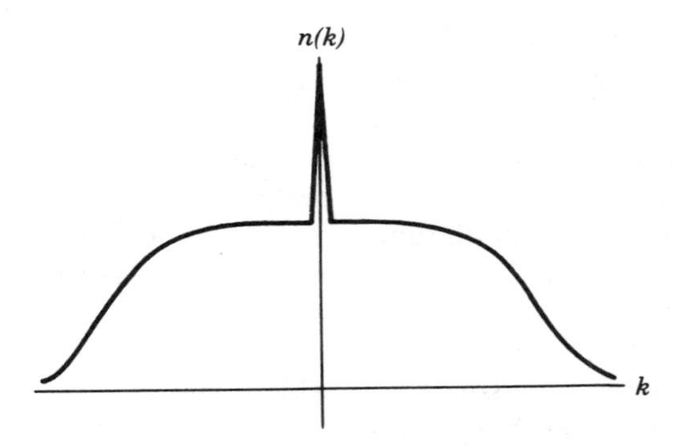

Figure 5.7. A momentum distribution with many particles with zero momentum.

that are not present in a boson calculation. For example, Ψ_T is occasionally equal to 0, which causes $E(\mathbf{R})$ to become infinite. Despite these and other difficulties, variational energies for a fermion system have been calculated by Monte Carlo techniques.[16,17]

In Chapter 8, another method, Green's function Monte Carlo, is described for solving the Schrödinger equation in many dimensions.

REFERENCES

1. J. P. Valleau and S. G. Whittington, A Guide to Monte Carlo for Statistical Mechanics: 1. Highways, in *Statistical Mechanics, Part A: Equilibrium Techniques*, Modern Theoretical Chemistry Series, Vol. 5, B. Berne, Ed., Plenum, New York, 1976.
2. J. A. Barker, Interatomic Potentials for Inert Gases from Experimental Data, in *Rare Gas Solids*, M. L. Klein and J. A. Verables, Eds., Academic Press, New York, 1976, Chap. 4.
3. N. Metropolis, A. W. Rosenbluth, M. N. Rosenbluth, A. H., Teller and E. Teller, *J. Chem. Phys*, **21**, 1087, 1953.
4. J. P. Valleau and D. N. Card, Monte Carlo estimation of the free energy by multistage sampling, *J. Chem. Phys.*, **57**, 5457, 1972.
5. G. Torrie and J. P. Valleau, Monte Carlo free energy estimates using non-Boltzmann sampling, *Chem. Phys. Lett.*, **28**, 578, 1974.
6. J. J. Erpenbeck and W. W. Wood, Statistical Mechanics of Time Dependent Processes, in *Statistical Mechanics, Part B: Time-dependent Processes*, Modern Theoretical Chemistry Series, B. J. Berne, Ed., Vol. 6, Plenum, New York, 1977.
7. R. J. Baxter, *Exactly Solved Models in Statistical Mechanics*, Academic Press, New York, 1982, pp. 14–23.
8. R. J. Baxter, *Exactly Solved Models in Statistical Mechanics*, Academic Press, New York, 1982, Chap. 7.
9. J. K. Lee, J. A. Barker, and F. F. Abrahams, Theory and Monte Carlo simulation of physical clusters in the imperfect vapor, *J. Chem. Phys.*, **58**, 3166, 1973.
10. A. J. C. Ladd, Monte Carlo simulation of water, *Mol. Phys.* **33**, 1039, 1977.
11. C. Zannoni, Computer simulations, in *The Molecular Physics of Liquid Crystals*, G. R. Luckhurst and G. W. Gray, Eds., Academic Press, New York, 1979, Chap. 9.
12. D. M. Ceperley and M. H. Kalos, Quantum Many-Body Problems, in *Monte Carlo Methods in Statistical Physics*, K. Binder, Ed. Springer-Verlag, Berlin, 1979, Chap. 4.
13. W. L. McMillan, Ground state of liquid ^{4}He, *Phys. Rev.*, **138**, A442, 1965.

14. R. D. Murphy and R. O. Watts, Ground state of liquid ^{4}He, *J. Low Temp. Phys.*, **2**, 507, 1970.
15. V. F. Sears, Kinetic energy and condensate fraction of superfluid ^{4}He, *Phys. Rev. B*, **28**, 5109, 1983.
16. D. M. Ceperley, G. V. Chester, and M. H. Kalos, Monte Carlo simulation of a many fermion study, *Phys. Rev. B*, **16**, 3081, 1977.
17. J. W. Moskowitz and M. H. Kalos, A new look at correlations in atomic and molecular systems. I. Application of fermion Monte Carlo variational method, *Int. J. Quan. Chem.*, **20**, 1107, 1981.

GENERAL REFERENCES

K. Binder, Ed., *Monte Carlo Methods in Statistical Physics*, Springer-Verlag, Berlin, 1979.

K. Binder, Ed., *Applications of the Monte Carlo Method in Statistical Physics*, Springer-Verlag, Berlin, 1984.

M. H. Kalos, Ed., *Monte Carlo Methods in Quantum Problems*, D. Reidel, Dordrecht, 1984.

6 SIMULATIONS OF STOCHASTIC SYSTEMS: RADIATION TRANSPORT

Here we shall consider Monte Carlo calculations in which modeling of an inherently stochastic system is carried out by artificial random sampling. Such simulations are widely applicable to many fields; for example, a computer operating system may be tested by giving it a randomly generated operating environment. The behavior of the central processor or the central memory may also be modeled by Monte Carlo. Areas of physics in which Monte Carlo simulations are often employed are radiation transport,[1,2] heat transfer,[3] and nuclear theory.[4] Reliability theory is a field in which simulations have recently been applied.[5,6] Imagine a system of interconnected objects; what problems occur when individual parts are partially or totally failing? This may be studied by simulating partial failure of the parts randomly and determining whether the whole system will fail. From this information, probability distributions for total failures and partial failures may be obtained. The analysis in reliability theory tends to be very complicated since the parts may not be connected by fault trees; therefore Monte Carlo will be a very convenient method to use. Importance sampling will be needed in the simulation to find the probability of unlikely events. Another area in which Monte Carlo simulation is frequently applied is queuing theory.[7,8]

There are several characteristics that are necessary to all well-designed simulations. A clear statement of the system to be simulated is needed. Once this is accomplished, the probability distribution functions that will be involved must be identified and explicitly defined. Methods must then be established to sample all the required pdf's. And finally, one must be able to understand how to interpret the information provided by the simulation. Someone who has designed and carried through one simulation is capable of doing any simulation. No matter what the field, the technical parts stay the same; it is just the jargon and the data that change.

In this chapter and the next we shall concentrate on simulations in just one area—the transport of radiation. This is a well researched field[1,2] where much work has been done. The simulations use many of the techniques described previously in this book; it will be apparent that these techniques are applicable in other fields as well. For a discussion of simulations in other areas, see Ref. 8.

6.1. RADIATION TRANSPORT AS A STOCHASTIC PROCESS

The transport of radiation is a natural stochastic process that is amenable to Monte Carlo modeling. To do so, it is not necessary even to write down the equations that are actually being solved. The particular simulation that we shall discuss in detail will be a simplification of the naturally occurring process, and we shall first outline the assumptions and limitations of our model. We shall restrict ourselves to considering only neutral radiation such as light, X-rays, neutrons, and neutrinos. The inclusion of charged particles such as electrons, protons, and alpha particles (nuclei of the ^{4}He atom) leads to a more complicated simulation since the radiation no longer travels straight-line paths between well-separated collisions. The details of the interaction of the radiation with the medium will be greatly simplified, and we shall neglect the effect the radiation may have on the medium. For example, neutrons change most media through transmutation of the nuclei. For this reason, the composition of a nuclear reactor slowly changes with time, which will affect the paths of later neutrons. A feedback mechanism is established that makes the equations describing the radiation transport nonlinear. Monte Carlo methods are not generally effective for nonlinear problems mainly because expectations are linear in character. A nonlinear problem must usually be linearized in order to use Monte Carlo techniques. In most cases the effect of transmutation is slow and the linearization is a good approximation.

We shall treat explicitly a simple geometry in discussing a simulation, but we shall sketch how more complicated geometries may be handled. Neutral radiation such as neutrons can be polarized since the neutron spin can have two different orientations, described as spin up or spin down. Neutrons scatter differently depending on their polarization, but this is not an important effect and will be ignored here. Finally, we shall neglect crystal and similar effects on the transport of radiation. When the radiation has a wavelength equal to the crystal spacing, the radiation waves will diffract and the pdf's for scattering will change.

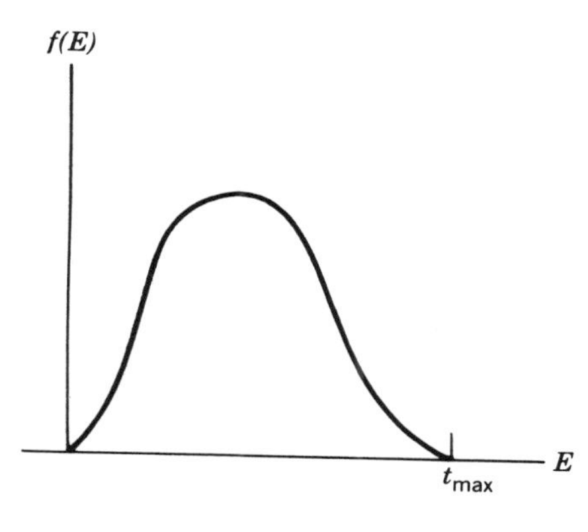

Figure 6.1. The energy distribution of the X-ray photons.

For the purposes of our discussion, we shall develop a simulation that models a familiar situation: the behavior of X-ray photons that are involved in a dental X-ray examination. These photons are initially produced by rapid deacceleration of an electron beam. The energy distribution of the X-rays varies from some lower limit up to the highest energy of the electron beam, as shown in Figure 6.1. The energy of any individual photon cannot be predicted. The electron beam is focused on a target from which the X-rays are emitted so the source is somewhat but not perfectly localized in space (Figure 6.2). The direction of the X-rays is, to some extent, random. The electron beam is turned on and off as the X-rays are needed so that they are not produced steadily in time.

The independent variables in the system detailed above are the energy, position, direction, and time of production of the photons. Each independent variable is random and may be described by a pdf. As X-rays move through a vacuum the spread in their direction makes their density fall (asymptotically) proportional to R^{-2}. Once in matter, the photon may interact with an individual atom; whether an interaction takes place and with which atom is a random process. The interaction usually involves the photons exchanging energy with an electron in the atom, effectively being scattered with reduced energy and changed

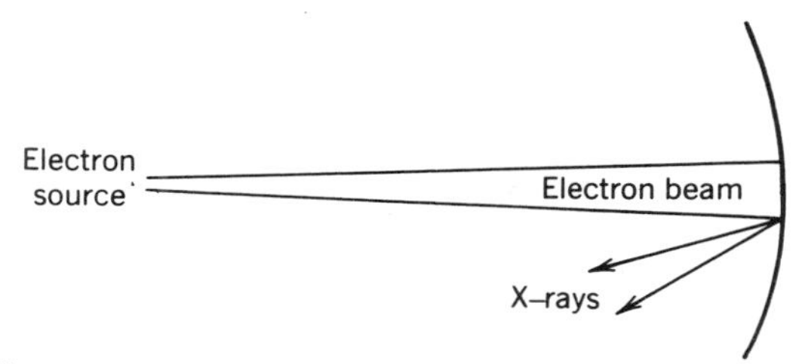

Figure 6.2. The collision of an electron beam on a target.

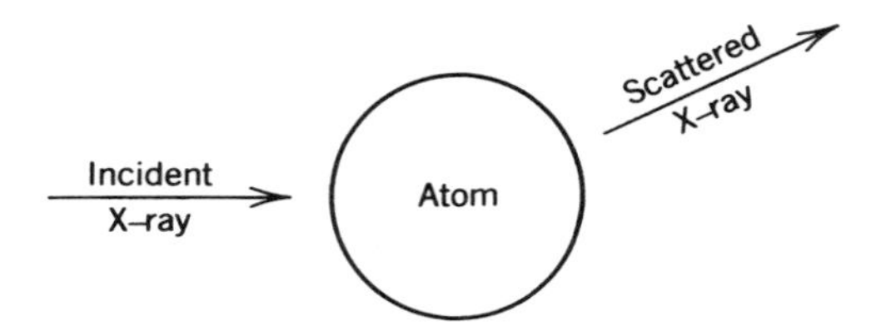

Figure 6.3. An X-ray scattering upon interaction with an atom.

direction (Figure 6.3). The photon may also give up all its energy to an electron, in which case it is considered to be absorbed and the history terminates. A relative probability based on knowledge of the relative rates is assigned to each event, scattering or absorption. An X-ray photon travels in straight lines until it interacts with an atom, at which point it may disappear or scatter at random. The scattered X-ray is subject to the same processes as an original X-ray of the same energy. In this and any other simulation, we must posit the pdf's governing each stochastic process that occurs. The distributions are deduced from theoretical and measured properties of X-rays. In some cases, educated guesses are used to fill in gaps in the exact form of the density functions.

The life history of a photon is followed until we lose interest in it or it is terminated. Termination occurs when a photon is absorbed by an atom. We lost interest in X-rays that become too low in energy to have substantial later effects or that move far enough away to have no appreciable chance of influencing the required answer. In summary, then, the process we wish to simulate is an X-ray photon born at a random position, direction, and time, which travels in straight line segments whose lengths are random. The photon interacts with the atoms constituting the medium at random, and its life history is concluded at random.

An appropriate description of the radiation must be chosen for the simulation. X-ray radiation can be considered an ensemble of X-ray photons and neutron radiation an ensemble of individual neutrons. The simulation can be performed by following the life history of a single photon or neutron. To specify completely the state of an X-ray photon, three variables are needed for the position, two or three for the direction, plus an energy variable and a time variable. Thus seven or eight independent variables are necessary.* In practical calculations three

*When the geometry is simpler, fewer variables may suffice. In slab geometry, one spatial coordinate and one direction variable are enough. In steady-state problems, time may be ignored.

variables are generally used for direction rather than, say, a polar and azimuthal angle minimally needed. The values that the independent variables can assume determine an ensemble of states of the system. How do we describe the states? Is it necessary to sample all the states in the ensemble in our simulation?

To answer the last question, we need to state clearly our objectives in doing the simulation. For example, with regard to dental X-rays, we can ask what dose of radiation is received at a particular position upon turning on the X-rays at a prescribed voltage and for a known duration of time. Another possible objective in simulating X-ray transport would be to determine the fraction of radiation that passes through a slab of matter of known thickness d (Figure 6.4). A modeling of nuclear radiation might try to predict the number of neutrons produced within a reactor compared with the number introduced by an external source. A simulation of a laser beam passing through dust could be used to predict how much of the light is scattered and detected elsewhere. The information that is sought from the simulation will indicate which states of the system must be sampled.

As stated above, seven or eight variables may be used to specify uniquely the state of the radiation. The position coordinates may be taken as the cartesian coordinates (x, y, z). The direction of the radiation can be specified by the direction cosines, $(\Omega_x, \Omega_y, \Omega_z)$. If the radiation is a beam of neutrons, the direction cosines can be calculated from the particle velocities, for example,

$$\Omega_x = v_x/(v_x^2 + v_y^2 + v_z^2)^{1/2}.$$

An energy or equivalent variable must be given. In scattering visible light by an atom, the energy of the photon hardly changes. Also, in dental X-rays, it is often the total dosage of radiation that is of interest, not the dosage as a function of time. This assumes, very reasonably, that

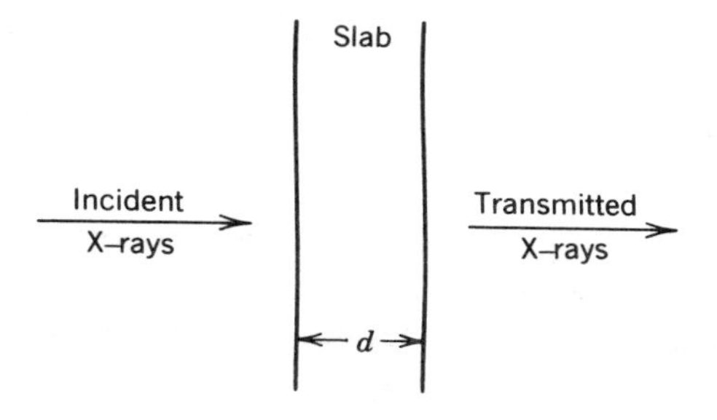

Figure 6.4. X-rays incident upon a slab.

the patient or technician does not move far as the photons traverse his or her body.

Once the system to be simulated has been clearly defined and the type of information wanted from the simulation is known, the structure of an appropriate code can be devised. For the simulation of dental X-rays, a possible structure is as follows:

1. Pick a set of source variables (initial state of system).
2. Follow the X-ray until it interacts with an atom.
3. Determine whether the X-ray scatters.
   a. If so, repeat from step 2.
   b. If not, terminate the history.

Steps 2 and 3 are repeated until the X-ray photon is absorbed or is no longer capable of affecting the answer to any appreciable extent.

4. Repeat the whole process from step 1 as many times as necessary to achieve the accuracy needed for the solution.
5. Take arithmetic average of answers of all the histories.

6.2. CHARACTERIZATION OF THE SOURCE

To realize step 1, sampling schemes must be developed for selecting the initial values of the independent variables from a source probability density function. In general, there exists a pdf for the production of radiation per unit volume, per unit direction, per unit energy, and per unit time:

$$S(x, y, z, \Omega_x, \Omega_y, \Omega_z, E, t).$$

We shall assume here that the source pdf can be factored into a product of pdf's, that is, position, direction, energy, and time are independent of one another:

$$S = S_x(x, y, z) S_\Omega(\mathbf{\Omega}) S_E(E) S_t(t). \tag{6.1}$$

This assumption simplifies our discussion; there are, however, important problems where factorization is not possible. In our simulation we shall assume that the X-rays are produced by what is termed a point source; that is, the X-rays are created at the spatial point (x_0, y_0, z_0). We assume (incorrectly) that each photon has the same energy E_0. The variables (x, y, z) and E are then perfectly determined. The appropriate pdf's are

$$S_x = \delta(x - x_0)\delta(y - y_0)\delta(z - z_0)$$

and

$$S_E = \delta(E - E_0). \tag{6.2}$$

The X-rays are produced by turning on the electron beam at t_1 and turning it off at t_2; this results in a pulse of X-rays in the time interval $(t_2 - t_1)$ (Figure 6.5). We shall assume it to be constant during that period. The corresponding pdf is

$$S_t(t) = \begin{cases} \dfrac{1}{t_2 - t_1}, & t_1 < t < t_2 \\ 0, & \text{otherwise.} \end{cases} \tag{6.3}$$

We take the intensity of X-rays to be uniform in all possible directions, so $S_\Omega(\mathbf{\Omega}) = 1/4\pi$. The FORTRAN coding that will sample the initial values of the system's independent variables for a point source is

```
X = X0
Y = Y0
Z = Z0
E = E0
T0 = T2 + (T2 - T1)*RN(D)
PHI = 6.28319*RN(D)
OMEGAZ = 1. - 2.*RN(D)
OMEGAX = SQRT(1. - OMEGAZ**2)*COS(PHI)
OMEGAY = OMEGAX*TAN(PHI)
```

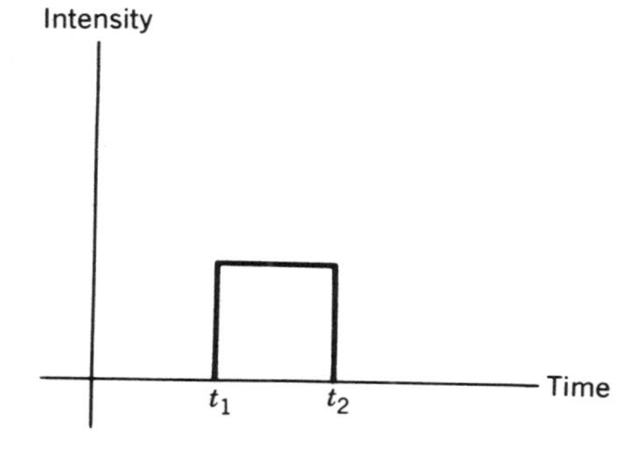

Figure 6.5. X-ray intensity as a function of time.

After some experience in simulation, one can code the selection of initial values without the necessity of writing down the source pdf's.

6.3. TRACING A PATH

The next step is to follow the path of a photon until it has an interaction. The essential feature here is that X-rays travel in straight lines until the interaction; so the change in position coordinates and time can be written down immediately as

$$x = x_0 + \Omega_x \cdot S,$$

$$y = y_0 + \Omega_y \cdot S,$$

$$z = z_0 + \Omega_z \cdot S,$$

$$t = t_0 + S/v,$$

where v is the radiation velocity and S is the distance to the next interaction with an atom. The probability per unit path length of having an interaction, Σ_T, is a property of the material and does not change with the distance the photon has traveled, at least to the point where the medium changes:

$\Sigma_T(\mathbf{X}, E)$ = probability per unit length along the X-ray path for any interaction.

As a consequence, the distribution of the distances, S, is exponential in homogeneous materials. The probability $U(S \mid \mathbf{\Omega}, E, \mathbf{X}_0)$ that the first interaction is at $S' > S$ is conditional upon the initial values of $\mathbf{\Omega}$, E, and $\mathbf{X}_0$ through the properties of the materials that are encountered in the path. The function $U(S \mid \mathbf{\Omega}, E, \mathbf{X}_0)$ is the complement of a cumulative distribution. For the moment, we shall ignore the conditional dependencies of U and consider only its behavior with respect to S. The value of U at some S_1 can be written

$$U(S_1) = U(S_2) + P\{S_2 \geq S > S_1\} \text{ for } S_2 > S_1, \tag{6.4}$$

and $P\{\cdots\}$ is the probability that an interaction occurred between S_1 and S_2. Equation (6.4) may be rewritten

$$U(S_1) - U(S_2) = U(S_1)P\{S_2 \geq S > S_1 \mid S > S_1\},$$

where $P\{\cdots\}$ is now the conditional probability that an interaction occurred if $S > S_1$. For small values of $S_2 - S_1$, the conditional probability that an event occurred is just the probability per unit length, $\Sigma_T(\mathbf{X}, E)$, multiplied by $(S_2 - S_1)$ plus higher-order terms:

$$U(S_1) - U(S_2) = U(S_1)\Sigma_T(S_1)(S_2 - S_1) + O(S_2 - S_1)^2. \tag{6.5}$$

Upon taking the derivative of Eq. (6.5) with respect to S_2, we find

$$-U'(S) = U(S)\Sigma_T(S)$$

(the subscript on S has been dropped) or

$$\frac{-U'(S)}{U(S)} = -\frac{d}{dS}\log U(S) = \Sigma_T(S). \tag{6.6}$$

Equation (6.6) can be integrated to yield the distribution function for S,

$$-\log U(S) + \log U(0) = \int_0^S \Sigma_T(S')\, dS'.$$

Since we know that exactly one interaction must take place for $S > 0$, $U(0) = 1$, and

$$U(S) = \exp\left[-\int_0^S \Sigma_T(S')\, dS'\right]. \tag{6.7}$$

In a homogeneous medium, Σ_T is independent of S', so that

$$U(S) = \exp[-\Sigma_T S]; \tag{6.8}$$

therefore the S's are indeed exponentially distributed.

Recall that $U(S)$ is the probability that an interaction takes place after a flight through a distance greater than S. We obtain a probability density function for the values of S by differentiating Eq. (6.8):

$$\frac{-dU(S)}{dS} = \exp\left[-\int_0^S \Sigma_T(S')\, dS'\right]\Sigma_T(S), \tag{6.9}$$

where the first factor on the right-hand side is the marginal probability that the path gets as far as S and the second term is the conditional probability that a collision occurs in dS at S. The distribution function is the probability of having a collision in a unit length along a ray at S.

For the purposes of the simulation, we need to decide how to sample the next event. As was discussed in Chapter 3, we can sample for a next event by equating a uniform random variable to a (cumulative) distribution function:

$$1 - U(S) = \xi' = 1 - \xi. \tag{6.10}$$

If the variable ξ is uniformly distributed, then $1 - \xi$ is also uniformly distributed and Eq. (6.10) becomes

$$U(S) = \xi = \exp\left[-\int_0^S \Sigma_T(S')\, dS'\right]$$

or

$$-\log \xi = +\int_0^S \Sigma_T(S')\, dS'. \tag{6.11}$$

For the case of a homogeneous medium, $\Sigma_T(S)$ is constant, so

$$-\log \xi = \Sigma_T S$$

and

$$S = -\log \xi / \Sigma_T. \tag{6.12}$$

We must sample the distance S from an exponential distribution. (The same idea, that an event has a fixed chance to occur per unit time or distance or whatever no matter how long you wait, occurs in many circumstances. The idea has frequent application in queuing theory.)

In most situations, the radiation travels through a medium that is a composite of several materials, each with its own value of Σ_T. Consider a ray and its direction in the medium as in Figure 6.6. To sample a next event, we must locate each distance to a boundary between two media, for example, $S_1, S_2, \ldots,$ with S being the total distance traversed. A uniform random number ξ is sampled and $-\log \xi$ is compared with $\Sigma_1 S_1$. If

$$-\log \xi < \Sigma_1 S_1, \quad \text{then } -\log \xi = \Sigma_1 S \quad \text{where } S < S_1;$$

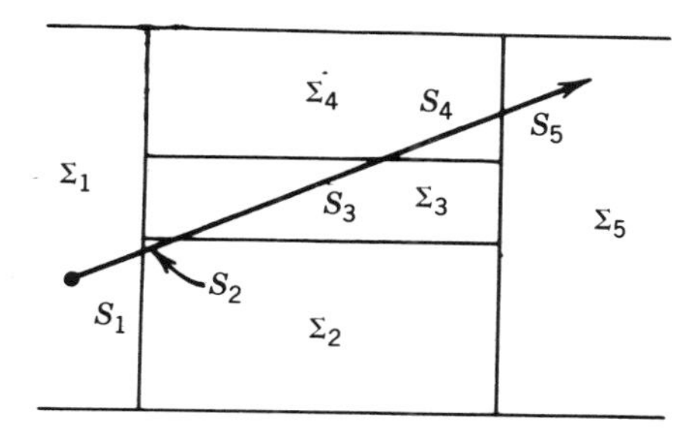

Figure 6.6. An X-ray traveling through a composite medium.

else test whether

$$\Sigma_1 S_1 < -\log \xi < \Sigma_1 S_1 + \Sigma_2 S_2; \quad -\log \xi = \Sigma_1 S_1 + \Sigma_2 (S - S_1),$$

where $S < S_1 + S_2$.

That is, we want to find the l such that

$$\sum_{j=1}^{l-1} \Sigma_j S_j \leq -\log \xi < \sum_{j=1}^{l} \Sigma_j S_j, \tag{6.13}$$

and then S is given by

$$S = \sum_{j=1}^{l-1} S_j - \left(\sum_{j=1}^{l-1} \Sigma_j S_j + \log \xi \right) \Big/ \Sigma_l, \tag{6.14}$$

where S_j is a partial distance in the jth medium encountered. The procedure is equivalent to finding that S where $-\log \xi$ intersects $\int_0^S \Sigma_T(S')\, dS'$ as shown in Figure 6.7. The procedure in Eq. (6.13) is easily programmed recursively.

By using the type of analysis just described it is very easy to sample for a next event in a complicated geometry. The major requirement is to be able to decide when a ray has intersected a surface. A quadric surface requires solving a quadratic equation; a linear surface requires solving a linear equation. The roundoff error in such calculations can become serious, so that the intersections will be erroneously determined. The programming of a complicated geometry can be simplified by taking advantage of repetitive geometries, for example, latticelike structures. The ability to describe almost any geometry leads some naive users to include every nut and bolt in the simulation. This is clearly not necessary,

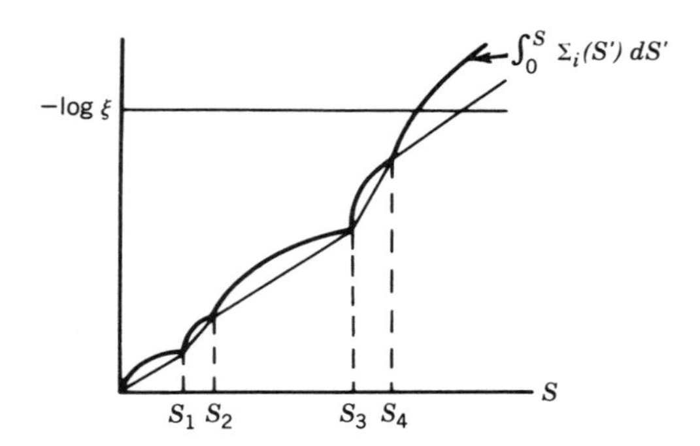

Figure 6.7. Finding the value of S where a next event will occur.

and the user must be sufficiently objective and experienced to be able to decide what is important.

Thus far we have discussed the X-ray source, the appropriate means to sample initial coordinates at random, and the straight-line transport of neutral radiation. As mentioned earlier, charged radiation does not travel in straight-line paths but continually changes its direction with interactions. An electron is deflected when it passes an atom and in addition it steadily loses energy, which leads to winding paths. Though the description of the transport of charged radiation is technically more complicated than that of neutral radiation, the nature of the simulation parallels what we are discussing here.[9]

6.4. MODELING COLLISION EVENTS

Eventually, the neutral radiation does interact with the atoms of the substance it is passing through. The exact outcome of the collision depends on the nature of the radiation; for example, a light photon will usually scatter, whereas X-ray and γ-ray photons will interact with the atom in a more significant way. To model the collision, the first step is to enumerate the possible events, then assign a probability to each class of events, and finally decide which events are important enough to be included in the simulation.

Suppose only two events can occur upon collision. The radiation can be absorbed by the atom, which will then either reemit some other radiation or convert the radiation energy to heat energy that is dissipated locally. For the moment, we suppose that our interest in the photon ceases upon absorption. The second event is the scattering of the radiation by the atom. The scattered radiation changes direction and loses energy, but for now we shall not include the latter effect explicitly in our discussion. Now that we have enumerated the possible events, we can assign a probability for scattering and absorption for the radiation at its particular energy in the medium. If

$$\xi < \text{prob}\{\text{absorption}\}, \qquad \text{absorption occurs;}$$

otherwise a scattering event occurs. If scattering happens, then the path of the radiation is continued as before but the direction of flight may be changed, allowing collisions to occur at random until the radiation is absorbed or leaves the area of interest.

Both neutrons and X-ray photons tend to scatter preferentially in the forward direction in most materials; the azimuth is uniformly distributed (for unpolarized radiation). The coordinates appropriate for the scattered X-rays are shown in Figure 6.8. The quantity $\mathbf{\Omega}$ is the unit vector in the old direction, θ is the scattering angle measured from $\mathbf{\Omega}$, and ϕ, the azimuthal angle, is chosen uniformly. In most simulations $\cos\theta$ is the natural variable for the pdf. As an example, consider a material in which scattering is isotropic in the forward hemisphere; the appropriate distribution function is

$$f(\cos\theta)\,d(\cos\theta) = \begin{cases} 2\cos\theta\,d(\cos\theta), & 1 > \cos\theta > 0 \\ 0, & \cos\theta < 0. \end{cases} \tag{6.15}$$

The new direction of the radiation is chosen by sampling $\cos\theta$ from Eq. (6.15) and then sampling $\phi = 2\pi\xi$. The old direction cosines were $(\Omega'_x, \Omega'_y, \Omega'_z)$; the new direction cosines are given by

$$\begin{aligned} \Omega_x &= \frac{\sin\theta}{\sqrt{1-\Omega_z'^2}}[\Omega'_y \sin\phi - \Omega'_z\Omega'_x \cos\phi] + \Omega'_x \cos\theta, \\ \Omega_y &= \frac{\sin\theta}{\sqrt{1-\Omega_z'^2}}[-\Omega'_x \sin\phi - \Omega'_z\Omega'_y \cos\phi] + \Omega'_y \cos\theta \\ \Omega_z &= \sin\theta\sqrt{1-\Omega_z'^2}\cos\phi + \Omega'_z \cos\theta. \end{aligned} \tag{6.16}$$

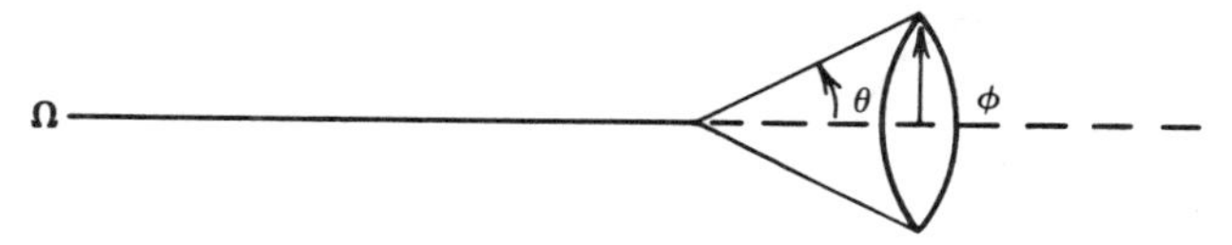

Figure 6.8. The coordinates for the scattered X-rays.

The set $(\Omega_x, \Omega_y, \Omega_z)$ is not unique—Eq. (6.16) results from a particular choice of the origin of ϕ—but does satisfy

$$\mathbf{\Omega} \cdot \mathbf{\Omega}' = \cos\theta \tag{6.17}$$

and

$$\Omega_x^2 + \Omega_y^2 + \Omega_z^2 = 1. \tag{6.18}$$

Equations (6.16) are not stable numerically, and the normalization given in (6.18) tends to drift from 1 after repeated usage of Eqs. (6.16). The $(\Omega_x, \Omega_y, \Omega_z)$ must be periodically renormalized. The von Neumann rejection technique (cf. Chapter 3) can be profitably used to choose $\sin\phi$ and $\cos\phi$, especially since they occur in association with a square root. If by chance $\Omega_z' = 1$, Eqs. (6.16) become indeterminant; this can be overcome by cyclic permutations of x, y, z in the equations. This change is also worth carrying out if Ω_z' is close enough to 1 to produce significant round-off error.

In addition to the two events discussed above, γ-rays can also undergo a process termed *Compton scattering*, in which a γ-ray photon interacts with an individual electron in the atom (Figure 6.9). The γ-ray is scattered, and the electron is ejected from the atom due to its increase in energy. Let E' be the energy of the γ-ray photon; then the *Compton wavelength* before scattering is

$$\lambda' = \frac{m_e c^2}{E'}, \tag{6.19}$$

where m_e is the electron rest mass. Following scattering, the Compton wavelength will be designated by λ. The joint probability of being scattered into $d\lambda$ and a solid angle $d\mathbf{\Omega}$ is

$$d\sigma = \frac{3}{16\pi}\sigma_T \left(\frac{\lambda'}{\lambda}\right)^2 \left[\frac{\lambda}{\lambda'} + \frac{\lambda'}{\lambda} + (\lambda' - \lambda)^2 - 2(\lambda - \lambda')\right] \cdot \delta(\cos\theta - 1 + \lambda - \lambda') \cdot d\mathbf{\Omega}\, d\lambda \tag{6.20}$$

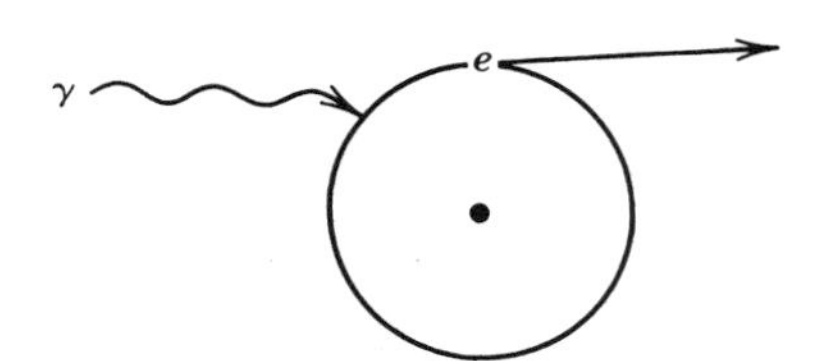

Figure 6.9. Compton scattering of γ-rays.

where σ_T is termed the Thomsen cross section. The quantity $\delta(\cos\theta - 1 + \lambda - \lambda')$ is the Dirac delta function, which couples $\mathbf{\Omega}$ and λ as required by the conservation of energy and momentum. The expression in Eq. (6.20) is the probability function for a γ-ray photon interacting with an electron and must be sampled in the simulation of γ-rays. The Dirac delta function requires that

$$\lambda - \lambda' = 1 - \cos\theta$$

or

$$\lambda = (1 - \cos\theta) + \lambda', \qquad \cos\theta \geq 0, \tag{6.21}$$

which is termed *Compton's law*. From Eq. (6.21) we see that $\lambda > \lambda'$, so $E < E'$ (cf. Eq. (6.19)) and the γ-ray loses energy.

REFERENCES

1. J. Spanier and E. M. Gelbard, *Monte Carlo Principles and Neutron Transport*, Addison-Wesley, Reading, Massachusetts, 1969.
2. L. L. Carter and E. D. Cashwell, Particle Transport Simulation with the Monte Carlo Method, ERDA Critical Review Series, TID-26607, 1975.
3. John R. Howell, Application of Monte Carlo to Heat Transfer Problems, in *Advances in Heat Transfer*, Vol. 5, T. F. Irvine, Jr., and J. P. Hartnett, Eds., Academic Press, New York, 1968.
4. Hugo W. Bertini, Spallation Reactions: Calculations, in *Spallation Nuclear Reactions and Their Applications*, B. S. P. Chen and M. Merker, Eds., Reidel, Dordrecht and Boston, 1976.
5. R. E. Barlow and F. Proschan, *Mathematical Theory of Reliability*, John Wiley and Sons, New York, 1965.
6. R. E. Barlow, J. B. Fussell, and N. D. Singpurwalla, Eds., *Reliability and Fault Tree Analysis*, Society for Industrial and Applied Mathematics, Philadelphia, 1975.
7. K. D. Tocher, *The Art of Simulation*, D. Van Nostrand, Princeton, New Jersey, 1963.
8. R. Y. Rubenstein, *Simulation and the Monte Carlo Method*, John Wiley and Sons, New York, 1981, Chapter 6 and references therein.
9. M. J. Berger, Diffusion of Fast Charged Particles, in *Methods in Computational Physics*, Vol. 1, B. Alder, S. Fernbach, and M. Rotenburg, Eds., Academic Press, New York, 1963.

GENERAL REFERENCES

G. Goertzel and M. H. Kalos, Monte Carlo Methods in Transport Problems, in *Progress in Nuclear Energy*, Vol. 2, Series I, Pergamon Press, New York, 1958, pp. 315–369.

J. Spanier and E. M. Gelbard, *Monte Carlo Principles and Neutron Transport*, Addison-Wesley, Reading, Massachusetts, 1969.

7 RANDOM WALKS AND INTEGRAL EQUATIONS

In Chapter 6 we discussed a highly simplified model of the transport of radiation and its simulation by a rather direct transcription of the natural stochastic processes into numerical (i.e., Monte Carlo) sampling procedures. We intend here to abstract a mathematical procedure that will relate the expected behavior of such simulations—which are a kind of random walk—to integral equations. This will have two consequences. One is that we shall be able to formulate an importance sampling procedure analogous to that of Section 4.1 that can be used to reduce the variance of Monte Carlo calculations of radiation transport. The second consequence will be that we shall recognize that certain classes of integral equations can be solved by the same simulation techniques.[1,2,3] Indeed, we shall be motivated to reformulate problems into integral equations so as to be able to exploit the analogy.

7.1. RANDOM WALKS

Let us now outline in the sparsest possible way the steps needed to carry out a Monte Carlo simulation of radiation transport.

1. Formulate a description of the sources of radiation. Interpret that description as a probability density function. Sample the probability density function to specify initial values of the coordinates in the simulation.
2. Formulate the tracing of a path and the description of interactions between elements of radiation and medium. Sample the probability distribution function for distance traveled and various probabilities to determine whether and what kind of radiation continues the process.

3. Repeat step 2 until either the radiation disappears (is absorbed) or becomes uninteresting.
4. During the iteration of step 2 count up interesting events to record physical results.

Steps 1–3 are, in effect, rules for carrying out a random walk of an object that moves from one point to another in a space of coordinates that describe the radiation. A minimal mathematical description of this requires four elements. First, we characterize the space on which the walk is defined. For present purposes, it can be $\mathbb{R}^n$; some readers will find it convenient to visualize the random walk on the real line.

Second, we need a description (i.e., a probability density function) of the "source." This is a function

$$\begin{gathered} S(X) \geq 0, \\ \int_{\Omega} S(X)\, dX = 1. \end{gathered} \tag{7.1}$$

The normalization is convenient and, as will be seen, implies no loss of generality.

The third element is a stochastic rule for moving from one point (say X') to another (call it X). This will be a density function $T(X \mid X')$ for sampling a new point X when the previous point of the walk was X'. Note that a similar kind of transition density was introduced in Section 3.7 in describing the $\mathrm{M(RT)}^2$ sampling method. We shall require

$$\begin{gathered} T(X \mid X') \geq 0, \\ \int T(X \mid X')\, dX \leq 1. \end{gathered} \tag{7.2}$$

That T is not normalized to 1 permits the possibility that the walk terminates [at X' with probability $1 - \int T(X \mid X')\, dX$].

The fourth element in our formulation is some variable of interest that we wish to know, at least conceptually. Our general quantity will be the density of arrivals at X, $\chi(X)$. That is, summing over all steps of the random walk and averaging over all possible walks, the expected number of times that a point is sampled within a region Ω is

$$\int_{\Omega} \chi(X)\, dX.$$

Recall that the walk starts with an X (call it X_0) sampled from $S(X_0)$. Then, if not absorbed, it moves to X_1 sampled from $T(X_1 \mid X_0)$. In general $T(X_n \mid X_{n-1})$ governs the nth move. The arrival at X can occur either because it was sampled from S or because it moved to X from an earlier point (call it Y). The total average density at X is the sum of these two:

$$\chi(X) = S(X) + \int T(X \mid Y)\chi(Y)\, dY. \tag{7.3}$$

The integral term on the right-hand side is the average density arrival at X from the next earlier arrival: $\chi(Y)\, dY$ is the chance that there was an arrival in dY; $T(X \mid Y)$ is the probability that this was followed by a move to X; one integrates over Y to average over all possible positions of the previous move.

Equation (7.3) is then an equation that describes the average behavior of the random walk. In a sense, since the outcome is a series of points $X_0, X_1, X_2, \ldots,$ the random walk can be regarded as a device for sampling the function χ that is the solution of Eq. (7.3). This is analogous to the procedure used by $M(RT)^2$, but there is one vital difference. In the latter the density function was prescribed. Now, χ is in general unknown. In $M(RT)^2$ we did not use a specific S, and the correctness of the sampling is true asymptotically. Here S is part of the description, and the correctness of the sampling requires that every X_n generated in the walk be used. Like earlier sampling procedures, the set $\{X_{nm}\}$ of points obtained from repetitions $m = 1, 2, \ldots, M$ of the whole random walk may be used to form estimators of integrals. That is, if

$$G = \int g(X)\chi(X)\, dX, \tag{7.4}$$

where χ is the solution of Eq. (7.3), then the quantity

$$G_M = \frac{1}{M} \sum_{m=1}^{M} \sum_{n=1}^{N} g(X_{nm}) \tag{7.5}$$

is an estimator for G in the (by now familiar) sense that

$$\langle G_M \rangle = G, \tag{7.6}$$

where the expectation implies averaging over all stochastic events that underlie the random walk.

7.2. THE BOLTZMANN EQUATION

Clearly the ideas of Section 7.1 apply to the random walk that describes successive observations of radiation passing through matter. We use position $\mathbf{X}$, energy E, direction $\mathbf{\Omega}$, and time t to specify completely the state of the photon or neutron or whatever and call $P = (\mathbf{X}, E, \mathbf{\Omega}, t)$ the point that characterizes that state. $\chi(P)$ will be the density with which radiation may be observed to emerge either from the source $S(P)$ or from a previous collision at P' followed by a flight from $\mathbf{X}'$ to $\mathbf{X}$ and a collision that changes E' to E and $\mathbf{\Omega}'$ to $\mathbf{\Omega}$. We shall call the transition density $K(P' \to P)$ in this case. Then the ideas introduced in Section 7.1 transcribe to yield the integral equation

$$\chi(P)\,dP = S(P)\,dP + \int K(P' \to P)\chi(P')\,dP'\,dP. \tag{7.7}$$

The first term is simply the source contribution at P. The second term is the average contribution from processes in which radiation emerging from a previous collision (or source) at P' arrives at P and emerges from collision there. The total next collision is, of course, averaged over all possible P' that can contribute at P. Equation (7.7) is the linear Boltzmann equation in integral form. The Monte Carlo simulation of radiation transport amounts to solving (7.7) by a random sampling method.

Any average quantity of interest that derives from radiation transport can be obtained by appropriate averages over $\chi(P)\,dP$. An expected value is defined as

$$G = \int g(P)\chi(P)\,dP, \tag{7.8}$$

where $g(P)$ is related to both the medium and the nature of the quantity to be determined. An estimator G_M may be formed as in Eq. (7.5), which permits Monte Carlo computation of G to be made using the steps P_{nm} of the numerical random walks. A subject of considerable theoretical and practical interest is the formulation of functions $g(P)$ that permit results of engineering interest to be computed efficiently. We omit discussion of this in this chapter except for one trivial example: if one wishes to determine that fraction of radiation that escapes into a vacuum from a convex medium, one has only to count the cases where that occurs; that is, $g(P) = 0$ when $\mathbf{X}$ is "inside" a medium and $g(P) = 1$ when $\mathbf{X}$ is "outside." The walk may be terminated on escape.

7.3. IMPORTANCE SAMPLING OF INTEGRAL EQUATIONS

In many problems of interest the chance that the random walk reaches that region where an answer is obtained is small. The prototypical case is that mentioned at the end of Section 7.2: one wishes to calculate the escape of radiation from a "thick" medium in which many collisions must have an unlikely favorable outcome for the walk to contribute to the estimator. That is, for a fraction of steps very close to 1, $g(P)$ is 0. We seek methods to influence the walk to make a favorable outcome rather likely.

The problem has an analogy to importance sampling as used to reduce the variance in Monte Carlo quadrature (cf. Section 4.1). If

$$G = \int g(X) f(X)\, dX$$

is to be calculated, sample $\tilde{f}(X)$ and rewrite

$$G = \int \frac{g(X) f(X)}{\tilde{f}(X)} \tilde{f}(X)\, dX.$$

A favorable $\tilde{f}$ will be one that is as nearly proportional to $|g(X)| f(X)$ as can technically be achieved. The Monte Carlo is improved when X is sampled so as to occur more often when the integrand is large, that is, when the contribution to the answer is large.

We shall employ a similar strategy here. First, we shall alter the density with which collisions are sampled in the hope of making them occur preferentially where contributions to the answer required are expected to be large. Second, we shall analyze what characterizes expected contributions. This is a more complicated question here since the random walk may have to make a number of steps in regions where $g(P) = 0$ before arriving at a region where $g(P) > 0$.

To achieve a modification of the collision density, we introduce an importance function $I(P) \geq 0$ and multiply Eq. (7.7) through by it. In the integral term we multiply and divide by $I(P')$ to obtain an integral equation for the product $I(P)\chi(P)$, which plays the role of an altered collision density.

Define S_0 by

$$S_0 = \int I(P'') S(P'')\, dP'';$$

then a new function $\tilde{\chi}(P)$ can be written in terms of S_0 and $I(P)$:

$$\begin{aligned}\tilde{\chi}(P) &\equiv \frac{I(P)\chi(P)}{S_0} \\ &= \frac{S(P)I(P)}{S_0} + \int K(P' \to P)\frac{I(P)}{I(P')}\frac{I(P')\chi(P')\,dP'}{S_0}. \end{aligned} \tag{7.9}$$

This equation has the same formal structure as Eq. (7.7), but the new source term is

$$\tilde{S}(P) = \frac{S(P)I(P)}{S_0}, \tag{7.10}$$

which is normalized to 1. The modified kernel is

$$\tilde{K}(P' \to P) = \frac{K(P' \to P)I(P)}{I(P')}; \tag{7.11}$$

$I(P')$ must not vanish when $K(P' \to P)$ does not. If we use Eqs. (7.10) and (7.11), the integral equation in (7.9) becomes

$$\tilde{\chi}(P) = \tilde{S}(P) + \int \tilde{K}(P' \to P)\tilde{\chi}(P')\,dP', \tag{7.12}$$

and the definition of an expected value can be rewritten

$$G = \int g(P)\chi(P)\,dP = S_0 \int \frac{g(P)}{I(P)}\tilde{\chi}(P)\,dP \tag{7.13}$$

with the further requirement that $I(P)$ cannot vanish when $g(P)$ does not. Any linear functional that could be evaluated using Eq. (7.8) can be evaluated by means of Eq. (7.13). The importance function modifies the path the radiation travels since the path from P' to P is now chosen from $\tilde{S}$ and from $\tilde{K}(P' \to P)$. For example, if $I(P) > I(P')$, the path will be biased toward P.

Suppose that in performing a simulation through one history, the points $P_1, P_2, \ldots, P_l$ are encountered. Then the estimate corresponding to Eq. (7.13) is

$$G \cong S_0 \sum_{i=1}^{l} \frac{g(P_i)}{I(P_i)}. \tag{7.14}$$

Clearly the variance will be reduced if each history contributes approximately the same value to the estimate of G. In what follows, we show how, in principle, each history may give the same result. This is done by arranging that only one collision—the last—give a contribution (rather than all l as in (7.14)) and that the answer in that case be independent of the previous history.

Suppose we are given an element of radiation somewhere. What is its chance of contributing to the quantity of interest? Let $J(P)$ equal the expected score (contribution to G) associated with radiation emerging from a collision or the source at P. The value of $J(P)$ is large where the radiation can ultimately score and is small elsewhere. $J(P)$ is called the importance since it assesses the likelihood of radiation at P to contribute to the score. An integral equation for $J(P)$ consistent with the above is

$$J(P) = g(P) + \int K(P \to P')J(P')\, dP', \tag{7.15}$$

where $g(P)$ is the direct contribution to the score at P. In addition, the radiation at P can have further collisions that may lead to contributions to G. The integral on the right-hand side is the average contribution from all possible positions P' that can be reached in a single collision after a collision at P. Equation (7.15) is adjoint to Eq. (7.7) for $\chi(P)$, and together the two equations form a dual. One manifestation of the duality is obtained by integrating both equations;

$$\int \chi(P)J(P)\, dP = \int S(P)J(P)\, dP + \int\int J(P)K(P' \to P)\chi(P')\, dP'\, dP,$$

$$\int J(P)\chi(P)\, dP = \int \chi(P)g(P)\, dP + \int\int \chi(P)K(P \to P')J(P')\, dP'\, dP.$$

The two double integrals are equal since P' and P are variables of integration and can be interchanged; hence

$$\int \chi(P)g(P)\, dP = G = \int S(P)J(P)\, dP. \tag{7.16}$$

Either of the two integral equations in (7.16) can be used to calculate G.

In the Monte Carlo sampling that uses $K(P' \to P)$, the expected number of collisions following a collision at P' is

$$N(P') = \int K(P' \to P)\, dP. \tag{7.17}$$

$N(P')$ is not easy to calculate in complicated situations; however, weighting by $J(P)$ we obtain a simple identity. The expected number of next collisions weighted by $J(P)$ is

$$N_J(P') = \int \frac{J(P)K(P' \to P)}{J(P')}\, dP. \tag{7.18}$$

From Eq. (7.15)

$$\int J(P)K(P' \to P)\, dP = J(P') - g(P'),$$

so

$$N_J(P') = \frac{J(P') - g(P')}{J(P')} = 1 - \frac{g(P')}{J(P')}. \tag{7.19}$$

One has $0 \leq g(P') \geq J(P')$ (cf. Eq. (7.15)), so that

$$0 \leq N_J(P') \leq 1;$$

the expected number of next collisions is never greater than 1 when $\tilde{K}_J = K(P' \to P)J(P)/J(P')$ is used. $N_J(P')$ can be viewed as the probability of continuing the random walk and $g(P')/J(P')$ as the probability of stopping. If

$$\frac{g(P')}{J(P')} = \frac{\text{current score contribution}}{\text{all future scores}}$$

is close to 1, then there is no need to go on. We shall use $N_J(P')$ as the Monte Carlo probability for continuing the simulation from P', and the contribution to G will be calculated only when the walk stops (previously, we were estimating G at every collision). By scoring at every collision of the radiation, there is an indefinite number of scores in the random walk, and it is not possible to have a zero variance calculation (the number of scores is a random variable). Scoring when the history ends, however, means we score exactly once and we can have a true zero variance estimator. Suppose we replace $I(P)$ in Eq. (7.14) by $J(P)$,

$$G = S_0 \sum \frac{g(P)}{J(P)}, \tag{7.20}$$

and evaluate Eq. (7.20) with a P selected with probability $g(P)/J(P)$. Then the estimator is

$$S_0 \frac{g(P)}{J(P)} \cdot \left| \frac{g(P)}{J(P)} \right|^{-1} = S_0 = \int J(P)S(P)\, dP = G. \tag{7.21}$$

The last equality is Eq. (7.16). The estimate obtained from any history is G, independent of P and the whole preceding random walk, and is therefore a true zero variance estimator. It is necessary, unfortunately, that we already know the answer to the problem, $J(P)$, to evaluate Eq. (7.21). We can still attempt to approach a zero variance calculation by using functions $I(P)$ in the neighborhood of $J(P)$ and thereby reduce the variance of the Monte Carlo significantly. This idea is used extensively in radiation transport.

As a simple illustration of these ideas, consider radiation impinging upon a thick slab of material, as shown in Figure 7.1. What fraction of the incident radiation will emerge from the slab? The radiation is often observed empirically to decay nearly exponentially within the slab, so $\chi(z) \sim e^{-\mu_0 z}$ (Figure 7.2). If we assume that the important radiation is that traveling normal to the slab, then a reasonable and simple choice of importance function is

$$I(z) \propto e^{\mu_0 z}. \tag{7.22}$$

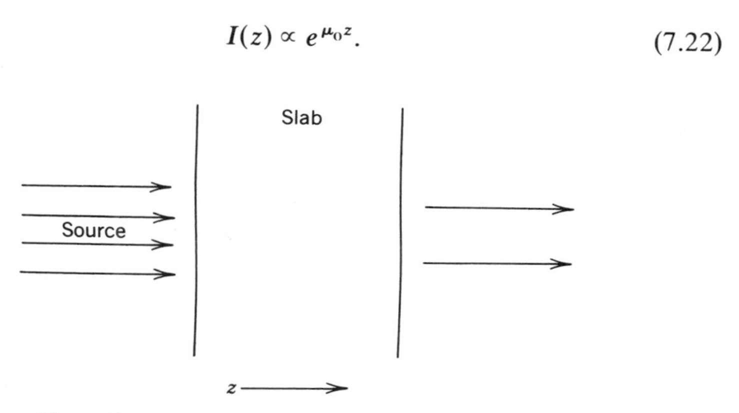

Figure 7.1. Radiation impinging on a slab of material.

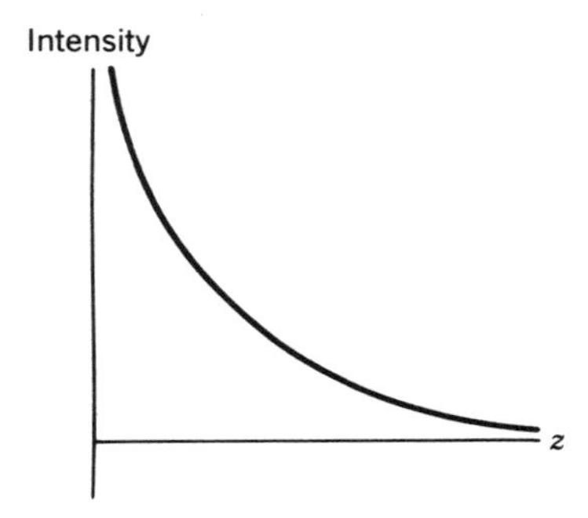

Figure 7.2. Radiation intensity within a slab.

This choice is correct only for radiation within the slab; its behavior beyond the slab is irrelevant since the random walk is terminated when the radiation exits. For radiation moving to the right in a direction making an angle with the normal whose cosine is ω,

$$K(z' \rightarrow z) = |\mu/\omega| \exp[-\mu(z - z')/\omega]. \tag{7.23}$$

Then (cf. Eq. (7.11))

$$\begin{aligned} \tilde{K}_I(z' \rightarrow z) &= K(z' \rightarrow z) I(z)/I(z') \\ &= |\mu/\omega| \exp[-(z - z')(\mu/\omega - \mu_0)], \end{aligned} \tag{7.24}$$

which is also an exponential and may be sampled in ways analogous to those used to sample z from K. The behavior of the effective attenuation coefficient in Eq. (7.24),

$$\tilde{\mu} = \mu - \omega\mu_0, \tag{7.25}$$

as a function of ω is interesting. For $\omega = 1$ (radiation parallel to the slab normal), $\tilde{\mu} = \mu - \mu_0 < \mu$. Thus, for such paths, the effective attenuation is diminished and typical paths are "stretched" to encourage escape. For $\omega = -1$, $\tilde{\mu} = \mu + \mu_0 > \mu$. That is, for radiation moving back toward the source, paths are "shrunk" on the average to prevent their moving too far back and so diminishing their chances of contributing to the tally of escaping particles.

Note also that Eq. (7.24) does not, unlike Eq. (7.23), have the simple normalization such that $\int \tilde{K}_I(z)\, dz$ is necessarily less than 1. For radiation for which $\tilde{\mu} < \mu$, a branching process—one particle turns into more on the average—is appropriate, and is an additional way in which a higher

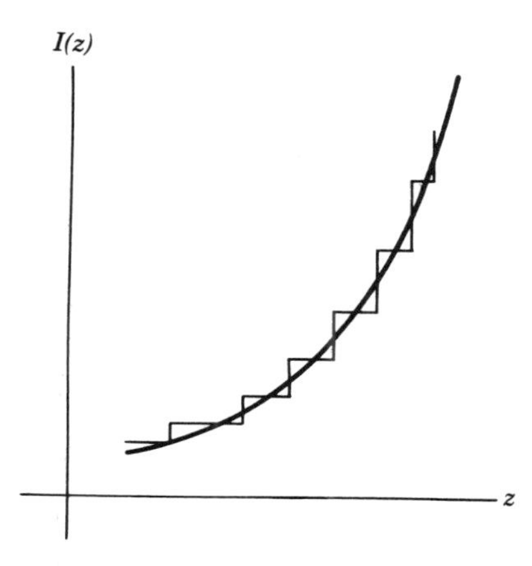

Figure 7.3. Approximating $I(z)$ by a step function.

score is encouraged for radiation already likely to escape. Correct calculation of the required normalization permits one to pass through $\tilde{\mu} = 0$ without any singular behavior of the sampling procedure.

For practical purposes, the exponential in $I(z)$ may be approximated by a step function, as shown in Figure 7.3, so $\tilde{K}_I(z)$ becomes the product of K and I, as shown in Figure 7.4, if $I(z)$ balances $K(z)$ on the whole. The sawtooth kernel shown in the figure may be sampled in a variety of ways, but in any case it is necessary to take account of the fact that $\int \tilde{K}_I(z)\, dz$ may exceed 1. It is best to introduce branching and so increase the chance of a score.

For a particle moving away from the slab exit, $\tilde{K}_I(z)$ is shown in Figure 7.5. The qualitative effect of I now reinforces that of K so that fewer particles are chosen for the continuation of the random walk and those that are chosen are closer, on the average, to z_0.

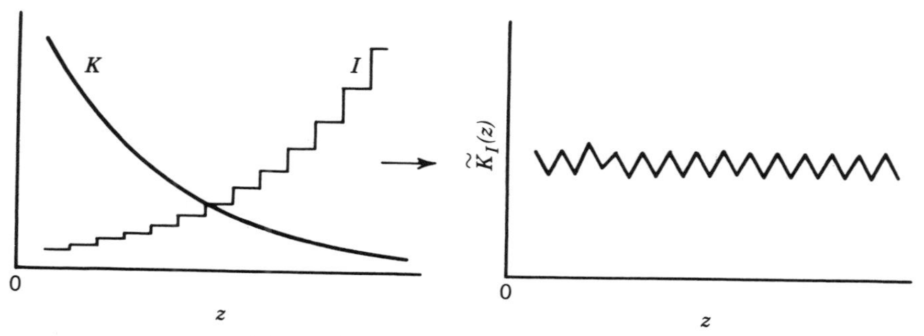

Figure 7.4. The product of $K(z)$ and $I(z)$.

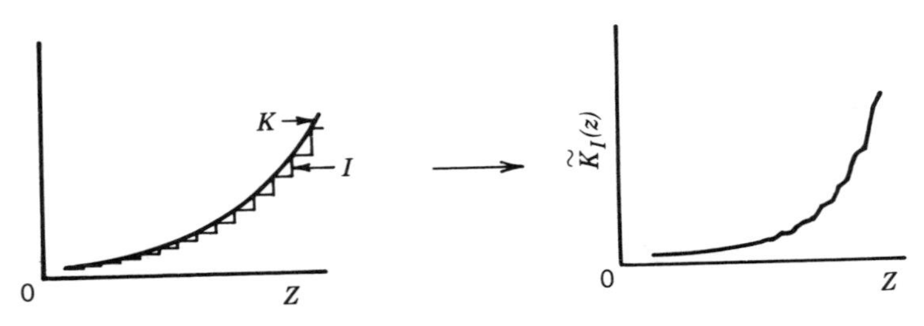

Figure 7.5. $\tilde{K}_I(z)$ for a particle moving backwards from the slab exit.

In general, the better that I approximates J and the sampling kernel approximates $I(P)K(P' \rightarrow P)/I(P')$, the more efficient the simulation is.

The treatment of random walks that simulate radiation transport and that therefore "solve" the linear Boltzmann equation in integral form is a model of how linear integral equations may be treated numerically by Monte Carlo methods.

REFERENCES

1. N. Metropolis and S. Ulam, The Monte Carlo method, *J. Amer. Stat. Assoc.*, **44**, 335 (1949)
2. M. D. Donsker and M. Kac, The Monte Carlo Method and Its Application, Proceedings, Seminar on Scientific Computation, November 1949, International Business Machines Corporation, New York, 1950, pp. 74–81.
3. N. Metropolis, in *Symposium on Monte Carlo Methods*, H. A. Meyer, Ed., John Wiley and Sons, New York, 1954, p. 29.

GENERAL REFERENCES

G. Goertzel and M. H. Kalos, Monte Carlo Methods in Transport Problems, in *Progress in Nuclear Energy*, Series I, Vol. 2, Pergamon Press, New York, 1958, pp. 315–369.

M. H. Kalos, Importance sampling in Monte Carlo shielding calculations, *Nuc. Sci. Eng.*, **16**, 227, 1963.

M. H. Kalos, Monte Carlo Solutions of Linear Transport Problems, in *Transport Theory*, Volume I, American Mathematical Society, Providence, 1969.

8 INTRODUCTION TO GREEN'S FUNCTION MONTE CARLO

In this chapter, we shall give an overview of the Green's function Monte Carlo (GFMC) method.[1] The power and importance of this method is that it provides a means by which the Schrödinger[2] equation and the Bloch[3] equation may be solved exactly for many-body boson[4] and also for fermion[5] systems. Since GFMC is a Monte Carlo method, the answers are subject to statistical sampling errors. For situations in which an exact answer to a quantum-mechanical problem is desired, the GFMC method is a useful tool.

The nature of the GFMC calculation and detailed discussions of the errors involved in such a calculation have been described elsewhere.[1–5] Here we shall develop a simple example to illustrate the method.

The essence of Green's function Monte Carlo can be described by four central ideas.

1. Monte Carlo methods can be used to solve integral equations. The most familiar application of Monte Carlo methods is to do many-dimensional quadratures; examples of such quadratures were given in previous chapters. As was shown in the example of the treatment of radiation transport (Chapter 7), however, Monte Carlo methods are equally useful in solving integral equations.

2. The Schrödinger equation and the Bloch equation can be transformed into integral equations by using Green's function for an appropriate operator. This Green's function, though, is usually not known explicitly (i.e., in closed form).

3. Random walks, which occur in a Monte Carlo calculation, can be constructed to generate the required Green's functions. For example, the operator in the Bloch equation, $-\nabla^2 + V + \partial/\partial t$, describes a diffusion process. The simulation of diffusion processes on computers is a well-known application of Monte Carlo methods. It is not surprising, there-

fore, that the generation of Green's functions by random walks can be done by a computer simulated stochastic process.

4. The statistical error inherent in a Monte Carlo calculation can be significantly reduced by introducing importance sampling. In principle, a calculation with importance sampling can have zero statistical error for a specific result (e.g., the energy of a quantum system).

8.1. MONTE CARLO SOLUTION OF HOMOGENEOUS INTEGRAL EQUATIONS

We shall now develop the GFMC method, stressing the four ideas outlined above. The Monte Carlo technique solves integrals and integral equations. Any arbitrary integral can be written in the form

$$G = \int_V f(x) g(x)\, dx,$$

where the function $f(x)$ has the properties that $f(x)$ is everywhere nonnegative and

$$\int f(x)\, dx = 1.$$

The latter requirement is just a matter of normalization. $f(x)$ is a probability density function, and we sample $f(x)$ in the course of the Monte Carlo calculation. The phrase "to sample $f(x)$" means an algorithm for producing a set $\{X_k\}$ such that for any $V_1 \subset V$,

$$\text{Prob}\{X_k \in V_1\} = \int_{V_1} f(x)\, dx. \tag{8.1}$$

The set $\{X_k\}$ is constructed by mapping uniform random variables, $\{\xi\}$, to $\{X_k\}$. Numerous examples of this process are given in earlier chapters. Here we shall assume that the set $X_1, X_2, \ldots, X_n$ has been generated so as to satisfy Eq. (8.1).

The arithmetic mean of $g(x)$ is given by

$$G_N = \frac{1}{N} \sum_{k=1}^{N} g(X_k). \tag{8.2}$$

The expectation value of G_N is

$$E(G_N) = \int f(x)g(x)\,dx. \tag{8.3}$$

From the central limit theorem of probability theory (Chapter 2),

$$G_N \cong G;$$

that is, G_N is a numerical approximation to the integral G. The approximation is better as the number of samples N becomes larger.

Suppose the set $\{Y_1, Y_2, \ldots, Y_m, \ldots, Y_n\}$ is drawn from the distribution $\chi(Y)$, and for each Y_m, another set $\{X_{m1}, X_{m2}, \ldots\}$ is drawn from $\lambda T(X \mid Y_m)$. The quantity λ is a known constant and T is an integrable, nonnegative function (which serves as a density function for X conditional on Y).

We shall call λT the transition probability (cf. Chapter 3). The average number of X_{mk} that are contained in V is

$$\{X_{mk} \in V\} = \int_V \int \lambda T(X \mid Y_m)\chi(Y_m)\,dY_m\,dX, \tag{8.4}$$

where V is a volume and

$$\lambda T(X \mid Y_m)\,dX = \text{average number of } X\text{'s in } dX,$$

$$\chi(Y_m)\,dY_m = \text{probability that } Y_m \in dY_m.$$

Suppose that λ is an eigenvalue of the integral equation

$$\chi(X) = \lambda \int T(X \mid Y)\chi(Y)\,dY. \tag{8.5}$$

Equation (8.5) can be solved by the iterative sequence

$$\chi^{(n+1)}(X) = \lambda \int T(X \mid Y)\chi^{(n)}(Y)\,dY. \tag{8.6}$$

This is called a *Neumann series*, and the sequence converges to the distribution $\chi(X)$ that is the eigenfunction of Eq. (8.5) having the smallest eigenvalue. Monte Carlo is used to iterate Eq. (8.6) and is especially worthwhile when X and Y are many dimensional. The trial function or initial iterate $\chi^{(0)}(X)$ may be chosen to be a set of uniformly distributed random numbers; a better guess can be obtained from approximate solutions of the problem. Variational calculations have proved especially useful (cf. section 5.2).

So far we have assumed that the eigenvalue is known but this is generally not the case. Let λ_T be an arbitrary (trial) value and λ denote the actual value.

If $\lambda_T > \lambda$, the population of points X grows, since the size of the kernel λ_T is made larger; and if $\lambda_T < \lambda$, the population of X decays. This can be seen from Eq. (8.4) since the expected number of points is proportional to λ. The choice of λ_T does not affect the convergence of the Neumann series to $\chi(Y)$ except insofar as the decay of the population may prevent the iteration from being continued. After the population converges in shape, it is generally desirable to set λ_T so as to maintain approximately constant population size.

8.2. THE SCHRÖDINGER EQUATION IN INTEGRAL FORM

We are interested in the ground-state energy and wave function ψ_0 of the Schrödinger equation

$$H\psi_0(\mathbf{R}) = E_0\psi_0(\mathbf{R}), \tag{8.7}$$

where H is the hamiltonian. The Green's function is defined by the equation*

$$HG(\mathbf{R}, \mathbf{R}') = \delta(\mathbf{R} - \mathbf{R}') \tag{8.8}$$

with the same boundary conditions on $G(\mathbf{R}, \mathbf{R}')$ as on $\psi_0(\mathbf{R})$ for both $\mathbf{R}$ and $\mathbf{R}'$. The formal solution for the Green's function is

$$G(\mathbf{R}, \mathbf{R}') = \sum_{\alpha} E_{\alpha}^{-1}\psi_{\alpha}(\mathbf{R})\psi_{\alpha}(\mathbf{R}'), \tag{8.9}$$

where the $\psi_{\alpha}(\mathbf{R})$ are the eigenfunctions of the problem. We can rewrite the Schrödinger equation in the form of Eq. (8.5),

$$\psi_0(\mathbf{R}) = E_0 \int G(\mathbf{R}, \mathbf{R}')\psi_0(\mathbf{R}')\, d\mathbf{R}', \tag{8.10}$$

*Green's function as defined in Eq. (8.8) may be either positive or negative. In the Monte Carlo calculation we shall be sampling $G(R, R')$ and therefore require that it be non-negative.

and solve for $\psi_0(\mathbf{R})$ by iteration,

$$\psi^{(n+1)}(\mathbf{R}) = E_T \int G(\mathbf{R}, \mathbf{R}')\psi^{(n)}(\mathbf{R}')\, d\mathbf{R}' \qquad (8.11)$$

where E_T is a trial energy. Suppose our initial distribution is $\psi^{(0)}(\mathbf{R})$. If we expand $\psi^{(0)}$ in the ψ_α,

$$\psi^{(0)}(\mathbf{R}) = \sum_\alpha C_\alpha \psi_\alpha(\mathbf{R}), \qquad (8.12)$$

then by using Eqs. (8.9) and (8.10) we find that

$$\psi^{(n)}(\mathbf{R}) = \sum_\alpha \left(\frac{E_T}{E_\alpha}\right)^n C_\alpha \psi_\alpha(\mathbf{R}). \qquad (8.13)$$

For sufficiently large n, the term $\alpha = 0$ dominates the sum in Eq. (8.13) and so

$$\psi^{(n)}(\mathbf{R}) \cong \left(\frac{E_T}{E_0}\right)^n C_0 \psi_0(\mathbf{R}) \qquad (8.14a)$$

and

$$\psi^{(n+1)}(\mathbf{R}) \cong \frac{E_T}{E_0}\, \psi^{(n)}(\mathbf{R}). \qquad (8.14b)$$

This result demonstrates that the iteration of Eq. (8.11) will eventually give the ground-state wave function. The ground-state energy E_0 can be obtained from Eq. (8.14b). It is worth reemphasizing that the Neumann series for solving an integral equation is applicable more generally. We are concerned with the use of Monte Carlo methods to perform the iterations.

The Green's function Monte Carlo method is easily demonstrated by using a simple example. Consider a particle on a line, enclosed by repulsive walls. The Schrödinger equation for this example is

$$\frac{d^2\psi(x)}{dx^2} = E\psi(x), \qquad \psi(x) = 0 \quad \text{for} \quad |x| \geq 1. \qquad (8.15)$$

The ground-state wave function and energy are

$$\psi_0(x) = \cos\left(\frac{\pi}{2}x\right), \qquad E_0 = \left(\frac{\pi}{2}\right)^2, \qquad (8.16)$$

and the appropriate Green's function is

$$G(x, x') = \begin{cases} \frac{1}{2}(1 - x')(1 + x), & x \le x' \\ \frac{1}{2}(1 + x')(1 - x), & x \ge x'. \end{cases} \tag{8.17}$$

So, for a particular choice of x, the Green's function is piecewise linear, and the Green's function may be sampled separately for each linear portion. The integral equation we solve in this example is

$$\psi(x) = E_T \int_{-1}^{1} G(x, x')\psi(x')\, dx'. \tag{8.18}$$

$G(x, x')$ may be identified with $T(X \mid Y)$ of Section 8.1 and used as a density. An iterative scheme can be described as follows. An initial population of points x' is chosen from $\psi^{(0)}(x')$; we shall assume it to be uniform. Then for each x'

1. construct the two triangles that form $G(x, x')$;
2. for each triangle compute $E_T A$, where A is the area of the triangle;
3. for each triangle, generate a random integer N, such that $\langle N \rangle = E_T A$ (i.e., N = largest integer smaller than $E_T A$ plus a uniform random number);
4. sample $\langle N \rangle$ points in each triangle to form the "next generation." The points are sampled from a triangular-shaped probability density normalized to 1.

8.3. GREEN'S FUNCTIONS FROM RANDOM WALKS

In our example we have assumed that we know Green's function for the problem. In general, however, especially in many-dimensional problems, we do not. We can exploit the relationship with diffusion processes to construct the Green's function in problems where it is not known. Consider again the particle in a box; we shall show, using a suitable expansion (realizable as a random walk), that the required Green's function may be sampled.

Let us choose as a zero-order Green's function an isosceles triangle. That is, for any x' in $(-1, 1)$, $G_T(x, x')$ will be an isosceles triangle extending from x' to the nearest endpoint and an equal distance on the other side. This introduces an extra and erroneous cusp at $x = x_1$, see Figure 8.1. The correct Green's function has cusps only at $x = x'$ and at

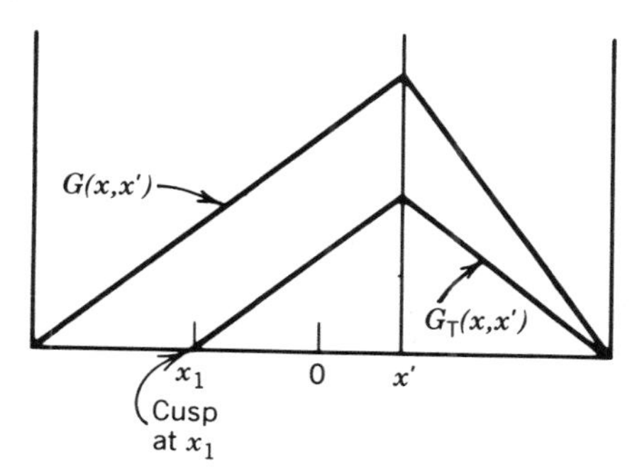

Figure 8.1. A $G_T(x, x')$ as an approximation to $G(x, x')$.

± 1. We correct for the improper choice of $G_T(x, x')$ by adding a triangle at x_1 whose central cusp cancels the extraneous one. Since we assume that we do not know the correct Green's function, we approximate the correction by another isosceles triangle centered at $x = x_1$. If it has the right size, it cancels the first erroneous cusp while introducing a new extraneous cusp that we must cancel. In so doing we define a (possibly) infinite sequence of iterations to sample $G(x, x')$ (Figure 8.2).

A random subsequence may be sampled by means of a random walk. Depending on the values of the x_j, the random walk need not be infinite. A graphic illustration of a random walk with three iteration steps is shown in Figure 8.3. Since x_2 is equal to 0, the third isosceles triangle does not create any extraneous cusps and the random walk terminates.

The process described above in the one-dimensional example is mathematically formulated as follows. Let Green's function be defined as

$$-\nabla^2 G(\mathbf{R}, \mathbf{R}_0) = \delta(\mathbf{R} - \mathbf{R}_0). \tag{8.19a}$$

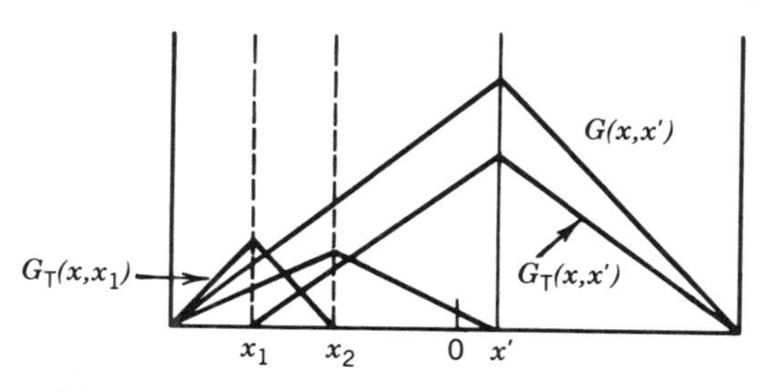

Figure 8.2. A sequence of functions $G_T(x, x_i)$ to sample $G(x, x')$.

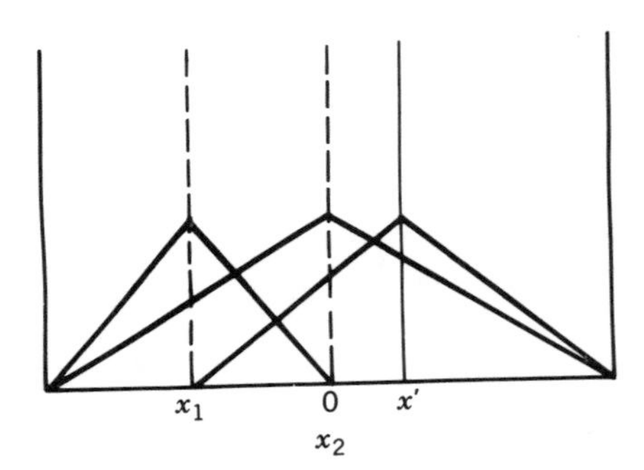

Figure 8.3. A random walk to sample $G(x, x')$ with three iterations.

It can be shown that the solution is symmetric:

$$G(\mathbf{R}, \mathbf{R}_0) = G(\mathbf{R}_0, \mathbf{R}). \tag{8.19b}$$

For every $\mathbf{R}_1 \in D$, construct a subdomain $D_u(\mathbf{R}_1) \subset D$ about $\mathbf{R}_1$ in such a way that the set of all D_u, $\{D_u\}$, covers the domain D. Within each D_u we shall define a Green's function $G_u(\mathbf{R}, \mathbf{R}_1)$ by

$$-\nabla^2 G_u(\mathbf{R}, \mathbf{R}_1) = \delta(\mathbf{R} - \mathbf{R}_1) \tag{8.20a}$$

and

$$G_u(\mathbf{R}, \mathbf{R}_1) = 0 \quad \text{for} \quad \mathbf{R}, \mathbf{R}_1 \notin D_u(\mathbf{R}_1). \tag{8.20b}$$

Now multiply Eq. (8.19a) by $G_u(\mathbf{R}_1, \mathbf{R}_0)$, multiply Eq. (8.20a) by $G(\mathbf{R}, \mathbf{R}_1)$, and subtract the two resulting equations. Integrate with respect to $\mathbf{R}_1 \in D_u(\mathbf{R}_0)$ and use Eq. (8.20b) to get

$$G(\mathbf{R}, \mathbf{R}_0) = G_u(\mathbf{R}, \mathbf{R}_0) + \int_{\partial D_u(\mathbf{R}_0)} [-\nabla_n G_u(\mathbf{R}_1, \mathbf{R}_0)] G(\mathbf{R}, \mathbf{R}_1)\, d\mathbf{R}_1. \tag{8.21}$$

where $-\nabla_n$ is the outward directed normal derivative for $\mathbf{R}_1$ on the boundary of $D_u(\mathbf{R}_0)$. Equation (8.21) is a linear integral equation for $G(\mathbf{R}, \mathbf{R}_0)$ in terms of the known subdomain Green's function $G_u(\mathbf{R}, \mathbf{R}_0)$. We may integrate it by means of a random walk, which will deliver a set of values of $\mathbf{R}$ drawn from $G(\mathbf{R}, \mathbf{R}_0)$ (cf. Section 7.1). In our one-dimensional example, the equation becomes

$$G(x, x_0) = G_u(x, x_0) + \left.\frac{\partial G_u(x, x_0)}{\partial x}\right|_{x=x_1} G(x, x_1) \tag{8.22}$$

and $\partial G_u(x, x_0)/\partial x$ is the probability that a next triangle will be constructed at x. When this probability is small, the random walk is likely to be terminated. The points in the next generation, $\{x\}$, of the iterative solution to the original integral equation, Eq. (8.18), are obtained by sampling $E_T \times G_u(x, x_1)$ for each point x_1 chosen in the random walk (Eq. (8.22)). Thus a random walk is generated for each x' in each generation that correctly samples $G(x, x')$ and evaluates Eq. (8.18).

8.4. THE IMPORTANCE SAMPLING TRANSFORMATION

The last technique that needs to be discussed in a general way is that of importance sampling. This is an example of the technique discussed earlier of transforming a problem into one that can be solved with lower—sometimes zero—Monte Carlo variance. In Chapter 4 the transformation appropriate for finite quadrature was introduced; in Chapter 7, the form required for variance reduction of integral equations and linear transport problems was discussed. In each case it proved useful to multiply the density function to be sampled for a point $\mathbf{R}$ by an estimate of the answer to be expected as a result of using $\mathbf{R}_0$ (i.e., the "importance" at $\mathbf{R}_0$). For the present case of a homogeneous integral equation, we may ask how to estimate that importance? A point not likely to produce future points may be said to be *unimportant*. We propose then to use the expected contribution to the population after many generations from a point at $\mathbf{R}_0$ as the importance. This is easy to compute by iterating from the trial function

$$\psi^{(0)}(\mathbf{R}) = \delta(\mathbf{R} - \mathbf{R}_0) = \sum_l \psi_l(\mathbf{R})\psi_l(\mathbf{R}_0), \tag{8.23}$$

where ψ_l are the eigenfunctions of the problem in Eq. (8.7). Calculating the population after n iterates of the integral equation of Eq. (8.11) gives

$$\psi^{(n)}(\mathbf{R}) = E_T^n \sum_l \frac{\psi_l(\mathbf{R})\psi_l(\mathbf{R}_0)}{E_l^n} \xrightarrow[n\to\infty]{} \left(\frac{E_T}{E_0}\right)^n \psi_0(\mathbf{R})\psi_0(\mathbf{R}_0). \tag{8.24}$$

The dependence of this on $\mathbf{R}$ reaffirms the convergence toward $\psi_0(\mathbf{R})$. The dependence on $\mathbf{R}_0$ shows that the importance of a point at $\mathbf{R}_0$; that is, its relative contribution to the population after many generations is $\psi_0(\mathbf{R}_0)$.

We should therefore hope that introducing an integral equation for $\psi_0(\mathbf{R})\psi(\mathbf{R})$ rather than for $\psi(\mathbf{R})$ might lead to lower variance in the population and hence in the energy estimate. That this anticipation is correct is seen as follows. Assuming $\psi_0(\mathbf{R})$ is a known function, one transforms Eq. (8.11) into

$$\tilde{\psi}(\mathbf{R}) \equiv \psi_0(\mathbf{R})\psi(\mathbf{R}) = E_{\mathrm{T}} \int \left[\frac{\psi_0(\mathbf{R})G(\mathbf{R},\mathbf{R}')}{\psi_0(\mathbf{R}')}\right] \tilde{\psi}(\mathbf{R}')\, d\mathbf{R}'. \qquad (8.25)$$

The nth iterate of this equation using $\delta(\mathbf{R}-\mathbf{R}_0)$ as initial function is again easily computed since the interior factors of ψ_0 in the iterated kernel of the last equation cancel in pairs, leaving

$$\tilde{\psi}^{(n)}(\mathbf{R}) = E_{\mathrm{T}}^n \psi_0(\mathbf{R}) \sum_l \frac{\psi_l(\mathbf{R})}{E_l^n} \to \left(\frac{E_{\mathrm{T}}}{E_0}\right)^n \psi_0^2(\mathbf{R}), \qquad (8.26)$$

independent of $\mathbf{R}_0$. In fact, the expected size of the population after n steps is

$$N_n = \int \tilde{\psi}^{(n)}(\mathbf{R})\, d\mathbf{R} = \left(\frac{E_{\mathrm{T}}}{E_0}\right)^n, \qquad (8.27)$$

also independent of $\mathbf{R}_0$. If $E_{\mathrm{T}} = E_0$, the expected population size is strictly constant with n as well. As we shall see later, it is possible to carry out the sampling (in the ideal case where $\psi_0(\mathbf{R})$ is known in advance) in such a way that the population size is precisely 1 at every stage. Thus the energy as estimated from the population growth has zero variance.

For the one-dimensional problem, that means that sampling point x given x' should be carried out by using

$$\frac{\pi^2}{4}\left[\cos\left(\frac{\pi}{2}x\right) G(x,x') \Big/ \cos\left(\frac{\pi}{2}x'\right)\right] \qquad (8.28)$$

as kernel. It is easy to verify that

$$\frac{\pi^2}{4}\int_{-1}^{1}\left[\cos\left(\frac{\pi}{2}x\right) G(x,x') \Big/ \cos\left(\frac{\pi}{2}x'\right)\right] dx = 1, \qquad (8.29)$$

independent of x', which, in this case, is the explicit guarantee that the population has no statistical fluctuations.

The qualitative effect of the transformation is easy to understand. As $x' \to \pm 1$ for the particle in a box, the expected next population $E_T \int G(x, x')\, dx \to 0$. By introducing $\cos(\pi x/2)$ as a factor in the density to be sampled, the number of points that turn up near $x = \pm 1$ is made small. At the same time, for those that do migrate near the wall, the contribution to the next generation is enhanced by the factor $\cos(\pi x'/2)$ that appears in the denominator of the modified kernel. The same balance will obtain for the case of a general potential energy function when the appropriate ψ_0 is used.

What has been established is that if ψ_0 is known and included as a *biasing factor* in the kernel and in the density to be sampled, it is possible to avoid most or all of the population size fluctuations. A practical procedure that suggests itself is to use some reasonable approximation to $\psi_0(\mathbf{R})$, which we shall call $\phi(\mathbf{R})$. Then introduce

$$\tilde{\psi}(\mathbf{R}) = \phi(\mathbf{R})\psi(\mathbf{R}) \tag{8.30}$$

and the transformed kernel is

$$\tilde{G}(\mathbf{R}, \mathbf{R}') = [\phi(\mathbf{R})G(\mathbf{R}, \mathbf{R}')/\phi(\mathbf{R}')]. \tag{8.31}$$

We expect that when $\phi(\mathbf{R})$ differs from $\psi_0(\mathbf{R})$ by a small amount, $\epsilon\phi_1(\mathbf{R})$, then the fluctuations in population and the variance of estimates of the energy will be $O(\epsilon^2)$. By now much experience has shown that physically appropriate approximate functions, especially those whose parameters have been chosen by minimizing the energy or by matching known observables, reduce the Monte Carlo variance by very large ratios. In principle the Monte Carlo calculation will give the same expected energy whatever the importance function $\phi(\mathbf{R})$ that is used. In fact, a very poor and unphysical choice of $\phi(\mathbf{R})$ can make the Monte Carlo convergence slower rather than faster, but this behavior, when it occurs, can usually be observed from the large fluctuations that accompany it. We note merely that appropriate avoidance of a hard wall or hard sphere potential is achieved by means of a trial function $\phi(\mathbf{R})$ that vanishes linearly in its neighborhood.

REFERENCES

1. M. H. Kalos, D. Levesque, and L. Verlet, Helium at zero temperature with hard-sphere and other forces, *Phys. Rev. A*, **9**, 2178, 1974.
2. D. M. Ceperley and M. H. Kalos, Quantum Many-Body Problems, *Monte*

Carlo Methods in Statistical Physics, Vol. 1, K. Binder, Ed., Springer-Verlag, Berlin, 1979.

3. P. A. Whitlock and M. H. Kalos, Monte Carlo calculation of the radial distribution function of quantum hard spheres at finite temperatures, *J. Comp. Phys.*, **30**, 361, 1979.
4. P. A. Whitlock, D. M. Ceperley, G. V. Chester, and M. H. Kalos, Properties of liquid and solid ^{4}He, *Phys. Rev. B*, **19**, 5598, 1979.
5. K. E. Schmidt and M. H. Kalos, Few- and Many-Body Problems, *Monte Carlo Methods in Statistical Physics*, Vol. 2, K. Binder, Ed., Springer-Verlag, Berlin, 1984.

GENERAL REFERENCES

D. Arnow, Stochastic Solutions to the Schrödinger Equation for Fermions, Courant Computer Science Report #23, Computer Science Department, New York University, 1982.

J. Carlson and M. H. Kalos, Mirror potentials and the fermion problem, *Phys. Rev. C*, **32**, 1735, 1985.

M. H. Kalos, Ed., *Monte Carlo Methods in Quantum Problems*, D. Reidel, Dordrecht, 1984.

APPENDIX

In much of the discussion of why and how to perform calculations using Monte Carlo methods, reference is made to a special type of random variable, the uniform random number ξ. This variable is expected to be distributed between (0, 1) with the property that

$$P\{\xi < x\} = x, \qquad x \in (0, 1). \tag{A.1}$$

Once we have a source of uniform random numbers, we can sample any other pdf and carry out Monte Carlo integrations and simulations. Thus it would appear that we must find such a source.

Since our interest lies in the use of Monte Carlo methods on computers, our source of uniform random variables must be capable of delivering them at the rate needed by a high-speed calculation. There are Monte Carlo calculations and some simulations in which the computation rate is dominated by the rate at which random numbers can be produced. Rejection techniques are often random number intensive. Clearly one should avoid a procedure in which random numbers take (say) a millisecond to produce on a machine that can do a million floating-point operations per second. Most natural sources of random events do not supply the events at such a rate (neglecting the difficulty of interfacing the source and the computer), and one would need an enormous table of previously compiled random numbers to complete a large calculation. Furthermore, we really want to be able to repeat our calculation at will, for debugging purposes and for various correlated sampling schemes. Thus for most calculational purposes, pseudorandom numbers (prn's) have been introduced. Pseudorandom numbers are sequences of numbers on (0, 1) that are easily generated on the computer and that will satisfy some statistical tests for randomness. Since the prn's are computed from a deterministic algorithm, they are not truly random and therefore cannot pass every possible statistical test.[1] The important criterion is that the prn generator used in a calculation satisfy those statistical tests that are most closely related to the problem being studied.[2–4] The use of test problems whose exact answers are known is also desirable.

A.1. MAJOR CLASSES OF prn GENERATORS

The earliest, purely deterministic method of generating pseudorandom numbers was the midsquares method proposed by von Neumann.[5] We take a number x_n that has $2a$ digits and square it. The resulting product has $4a$ digits, of which the a most significant and the a least significant digits are discarded. Then x_{n+1} is formed from the remaining $2a$ digits,

$$x_{n+1} = \left[\frac{x_n^2}{b^a}\right] - \left[\frac{x_n^2}{b^{3a}}\right] \cdot b^{2a}, \tag{A.2}$$

where b is the base of the number representation. This method of generating pseudorandom numbers was soon abandoned since it degenerates into a cyclic one, often with a very short period. In the worst case the period comprises just one number.

A.1.1. Linear Recurrence Methods

Most of the prn generators now in use are special cases of the relation

$$x_{n+1} \equiv a_0 x_n + a_1 x_{n-1} + \cdots + a_j x_{n-j} + b \pmod{P}. \tag{A.3}$$

One initiates the generator by starting with a vector of $j+1$ numbers $x_0, x_1, \ldots, x_j$. The generators are characterized by a period τ, which in the best case cannot exceed P^{j+1}. The length of τ and the statistical properties of the prn sequences depend on the values of the a_j, b, and P.

A.1.2. Multiplicative Congruential Generators

With the choice of $a_j = 0$, $j \geq 1$, and $b = 0$, one has the multiplicative congruential generator introduced by Lehmer,[6]

$$x_{n+1} \equiv \lambda \cdot x_n \pmod{P}. \tag{A.4}$$

In most applications P is chosen to be 2^β or 10^β, depending on whether the computer is binary or decimal, and β is the word length in bits. To obtain the maximum period τ for the generator, the parameter λ must be chosen to satisfy certain criteria.[7] But in addition to a large period, we want our generator to produce sequences of prn that are indistinguishable from truly random sequences in statistical tests. Due to the extensive work by Niederreiter[3,8] and Fishman and Moore[9] one can tailor the choice of λ (given P) to give excellent distributions of s-tuples of successive prns (i.e., vectors $(x_n, x_{n+1}, \ldots, x_{n+s-1})$ in s-dimensional space). In a famous paper, Marsaglia[10] showed that the multiplicative congruential generators then in use gave very poor distributions of s-tuples, and he questioned the use of such generators in serious Monte Carlo work. We can now, however, choose with confidence a set of parameters λ and P that will give us the desired statistical properties in many-dimensional spaces.

If the parameter b in Eq. (A.3) is not set equal to 0, then we have the mixed congruential generator

$$x_{n+1} \equiv \lambda x_n + b \pmod{P}. \tag{A.5}$$

The exact value of the increment b, provided that it is nonzero, has very little influence on the behavior of the generator; the statistical properties are mostly governed by the choice of λ.

A.1.3. Tausworthe Generators

The linear recurrence relation in Eq. (A.3) can be used to generate pseudorandom digits in addition to numbers. Tausworthe[11] introduced a method to produce binary digits; that is, P is 2. All the coefficients a_j and the x values are either 1 or 0. If the characteristic equation

$$1 + a_0 y + a_1 y^2 + \cdots + a_{n-1} y^{n-1} + y^n \tag{A.6}$$

is a primitive polynomial over the Galois field of two elements, then we get a sequence with maximum period $2^{n+1} - 1$. Recent work[12] has focused on the theoretical behavior of Tausworthe generators under various statistical tests. The aim is to identify the values of the a_j (a sequence of 0s and 1s) and n that will give statistical independence of successive terms in sequences.

One recently proposed generator[13] uses the primitive trinomial

$$1 + y^p + y^q, \qquad q > p. \tag{A.7}$$

A sequence of Q-bit (where Q is the word size of the computer) random integers $\{x_i\}$ are considered as m columns of random bits and a bitwise *exclusive or* (addition without carry), indicated by $\oplus$ is then performed,

$$x_k = x_{k-q+p} \oplus x_{k-q}\,. \tag{A.8}$$

A number between (0, 1) is computed from x_k/Q. The previous q iterates must be stored in this scheme. A suggested trinomial is $p = 103$, $q = 250$.

The initialization of the Tausworthe generator must be carried out carefully. If the ith and jth bits are identical in each of the first q integers, then they will remain identical in the whole sequence. One way to avoid this is to initialize the Tausworthe generator using integers produced by a good multiplicative congruential generator.

A.1.4. Combination Generators

For sophisticated Monte Carlo studies such as in geometric probability or estimating distribution functions, currently available prn generators may not be good enough. They fail stringent tests that mimic the behavior of the sophisticated Monte Carlo calculations. Marsaglia[14] proposes several new prn generators that are simple combinations of standard generators using computer operations +, −, *, and the bitwise exclusive or. The attractive features of the combination prn generators are that the resultant numbers tend to be more uniform and more nearly independent. Much longer periods are possible as well.

A.2. STATISTICAL TESTING OF prn GENERATORS

It is unlikely that any one prn generator will be appropriate in every application. There are requirements, however, that all good prn generators should meet. For example, the sequence of random numbers should be equidistributed in the interval (0, 1). To check the usefulness of a particular random number generator, it can be exercised by a battery of statistical tests to judge how well it mimics the behavior of a sequence of truly random numbers. We shall give some examples of the types of statistical tests used. A more thorough description is given in Refs. 2, 3, and 4.

A.2.1. Equidistribution Test

The interval (0, 1) is divided into k subintervals. The number N_j of values falling into each subinterval is determined for a sequence of prn,

$x_1, \ldots, x_N$. Then a chi-square test is performed where the expected number in each subinterval is N/k. The chi-squared value is then

$$\chi^2 = \frac{k}{N} \sum_{j=1}^{k} \left(N_j - \frac{N}{k} \right)^2 . \tag{A.9}$$

In a related test, Niederreiter[3] uses a "discrepancy" D_N, whose small size provides a valid measure of uniformity.

A.2.2. Serial Test

In this test one checks the interdependence between successive prn's in a sequence. We consider sets of points

$$\mathbf{x}_n = x_n, x_{n+1}, \ldots, x_{n+s-1}$$

in s-dimensional space, where $s \geq 2$. The space is divided into r^s partitions, and the frequency with which s-tuples fall into each partition is measured for a large sequence of prn's. Truly random numbers are uniformly distributed in the s-dimensional space, so we may compute a χ^2 value for our prn generator as

$$\chi^2 = \frac{r^s}{N} \sum^{r} \left(N_{j_1, \ldots, j_s} - \frac{N}{r^s} \right)^2 , \tag{A.10}$$

where $N_{j_1, \ldots, j_s}$ is the number of s-tuples falling in partition $((j_{i-1}/r), (j_i/r))$, $i = 1, \ldots, s$, $j_i = 1, \ldots, r$. χ^2 should behave asymptotically as a χ^2 distribution with $r^s - 1$ degrees of freedom. Another version of this test[3] calculates the discrepancy of the $\mathbf{x}_n$.

A.2.3. Runs-up and Runs-down Test

Here we compare the magnitude of a prn with the preceding one. If $x_{n-1} > x_n < x_{n+1} < x_{n+2} > x_{n+3}$, then we say we have a run-up of length 3. In a sequence of prn's we count the number of run-ups of length 1, length 2, ... For truly random numbers, where each of the x_i is independent of the others and all permutations are equally likely, we expect the number of occurrences of run-ups (or run-downs) of length r to be given by the formula

$$R_r = \frac{(N+1)(r^2+r-1) - (r+2)(r^2-r-1)}{(r+2)!} . \tag{A.11}$$

N is the total number of samples in the prn sequence. Since adjacent runs are not independent, we cannot apply a χ^2 test directly but must compute a different statistic[2]

$$V = \frac{1}{N} \sum_{1<i,j<t} (\text{Count}(i) - R_i)(\text{Count}(j) - R_j) a_{ij}. \qquad \text{(A.12)}$$

Count(i) is the actual observed number of run-ups (or run-downs) of length i, and the a_{ij} are the elements of the inverse of the covariance matrix. Once V is computed, it can be compared with a χ^2 distribution of degree t.

The run test is an important one. Linear congruential generators with multipliers that are too small tend to have runs lengthier than expected.

A.2.4. Testing a Random Number Generator

The tests described in Sections A.2.1–A.2.3 plus others[2–4] are used to judge the quality of a particular prn generator. For multiplicative congruential generators, it is now possible to choose parameters λ and P such that the behavior is known to mimic certain behaviors of truly random numbers very well.[8,15] For example, the generator

$$\begin{aligned} \lambda &= 1812433253, \\ P &= 2^{32}, \\ b &= \text{odd}, \end{aligned} \qquad \text{(A.13)}$$

gives excellent distributions of pairs of points on a 32-bit machine. We may ask how this prn generator behaves on other tests as well.

For the equidistribution test, we generated 10,000 random numbers at a time and placed them in 100 bins equispaced on (0, 1). The expected number in each bin is 100, and we computed a χ^2 value. The whole test was repeated 1000 times to generate 1000 values of χ^2. If the sequence being treated is uniform, then $\chi_1^2, \chi_2^2, \ldots, \chi_{1000}^2$ should behave like a random sample from a $\chi^2(99)$ distribution, and this can be tested by again applying a chi-square test.[2,4] We sort the values of χ_i^2, $i = 1, \ldots, 1000$ into 100 equally probable intervals $(0, \chi^2(99, 0.01))$, $(\chi^2(99, 0.01), \chi^2(99, 0.02)), \ldots, (\chi^2(99, 0.98), \chi^2(99, 0.99))$, $(\chi^2(99, 0.99), \chi^2(99, 1.0))$, where the expected number B_i in each interval is 10. Figure A.1 contains the distribution of χ^2 values obtained. The overall chi-squared value is

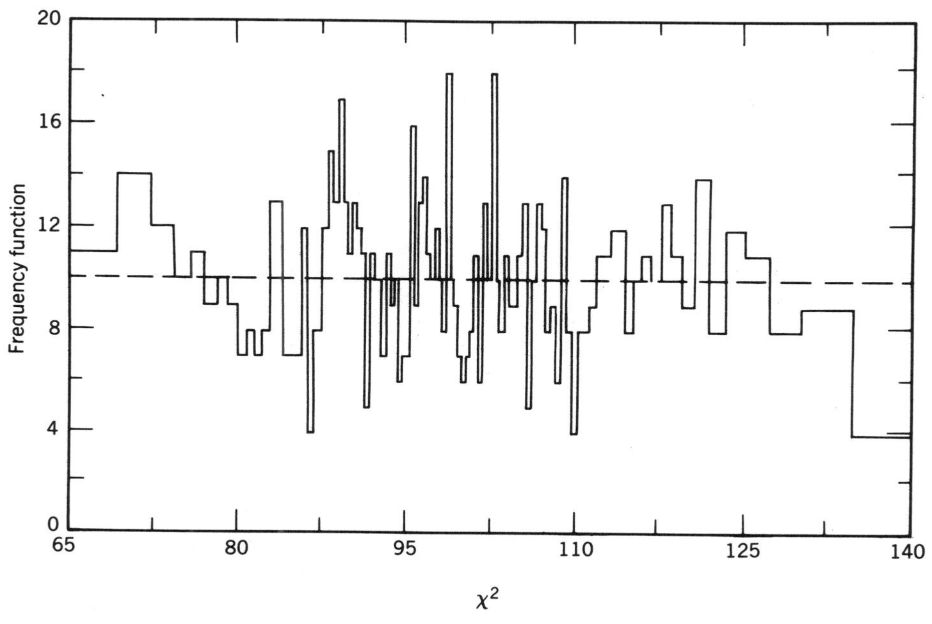

Figure A.1. The frequency function of observed values of χ^2 after 1000 samples. The bins are equally probable intervals where the expected number of χ^2 in each bin is 10. The pseudorandom number generator was Eq. (A.13).

then

$$\chi_T^2 = \sum_{j=1}^{100} \frac{(B_j - 10.)^2}{10.} = 82.2$$

for the random number generator in Eq. (A.13). In this case, the smaller the value of χ_T^2 the better the generator. Good, popular generators generally have values in the range 80–82.[4]

The serial test (Section A.2.2) was carried out for $s = 2$. That is, we tested the interdependence between pairs of prn's. We considered 10,000 pairs of prn's at a time and partitioned them in 10×10 array. A typical distribution of the pairs of prn's is shown in Table A.1 with its associated χ^2 value. Since we have 100 partitions, the expected number of pairs in each partition is 100. As in the equidistribution test above, we repeated the whole test 1000 times to generate 1000 χ^2 values and computed an overall χ_T^2. In this case, $\chi_T^2 = 104$, which is a reasonable value for a sample from a $\chi^2(99)$ distribution.

TABLE A.1. Partitioning of 10,000 Pairs of Pseudorandom Numbers Generated by Eq. (A.13) into a 10 × 10 Array

78	83	89	106	98	105	89	84	118	106
98	100	118	83	93	112	100	102	94	112
97	76	95	94	101	77	105	103	96	110
106	102	123	102	110	88	112	106	111	84
91	98	96	98	107	114	101	112	108	91
83	93	103	97	91	96	98	89	118	98
89	104	101	116	111	104	105	108	101	95
92	107	95	96	90	106	110	101	106	90
106	100	100	85	102	85	104	112	93	97
91	118	109	114	96	88	119	112	110	84

$\chi^2 = 105.2$

The pseudorandom number generator in Eq. (A.13) was also exercised by a runs-up test. A record of runs of length 1–5, 6 and greater was produced, and the statistic V in Eq. (A.12) was determined. Table A.2 shows the observed number of runs of length r and the expected number from Eq. (A.11). The value of V, 3.48, is to be compared with a $\chi^2(6)$ distribution and is an acceptable value. So for the few tests we have applied to generator Eq. (A.13), we have confidence that in calculations where one or a pair of prn's are needed at a time, Eq. (A.13) is a good source.

TABLE A.2. Observed Number of Runs of Length *r* Compared with the Expected Number after 10,000 Samples for Pseudorandom Number Generator (A.13)

Length of run	Observed	Expected[a]
1	1661	1667
2	2068	2083
3	889	916
4	289	263
5	60	58
6 and over	13	12
	$V = 3.48$	

[a]Calculated using Eq. (A.11).

A.2.5. A Bad Random Number Generator

Now that we have given an example of a good random number generator, we should like to contrast it with the behavior of a bad one. This time we use a congruential generator[2] with the following constants:

$$\lambda = 2^{18} + 1, \qquad x_0 = 314159265,$$
$$P = 2^{35}, \qquad b = 1, \tag{A.14}$$

where x_0 is the initial seed.

As in the development given above, we shall apply the equidistribution test and determine χ_T^2. In Figure A.2 is shown the distribution of the values of χ^2, which is seen to be skewed toward higher values. This is reflected in the value for χ_T^2, which is 1022.0, an unacceptably high value.

The poor behavior of generator (A.14) with the equidistribution test becomes worse in the serial test. Again, pairs of prn's were partitioned

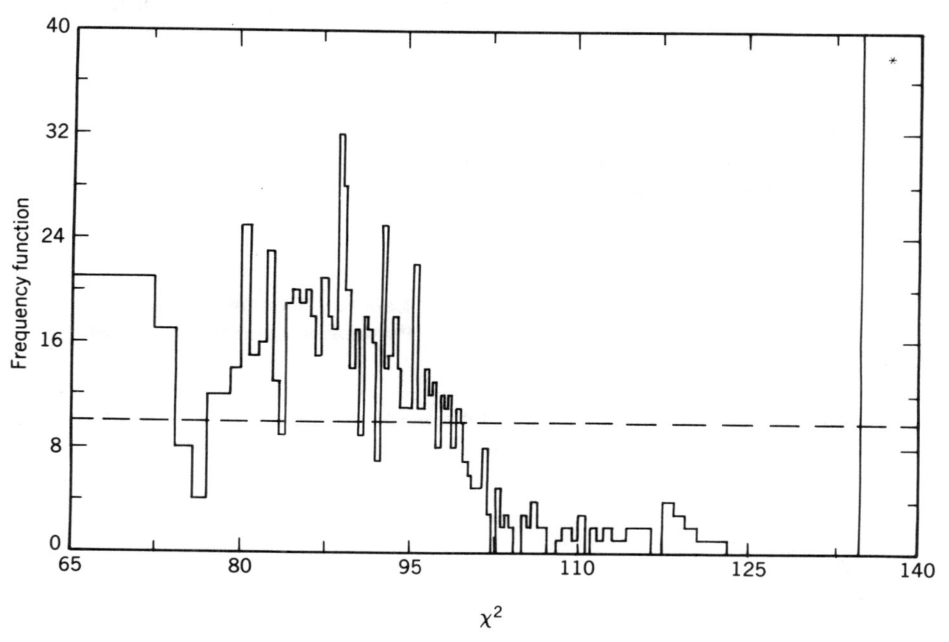

Figure A.2. The frequency function of observed values of χ^2 after 1000 samples using prn generator Eq. (A.14). The expected number of χ^2 in each bin is 10. The value of the bin indicated by (*) is offscale.

TABLE A.3. Partitioning of 10,000 Pairs of Pseudorandom Numbers Generated by Eq. (A.14) into a 10 × 10 Array

332	0	0	0	0	0	0	0	99	577
575	311	0	0	0	0	0	0	0	99
99	557	304	0	0	0	0	0	0	0
0	131	574	308	0	0	0	0	0	0
0	0	106	580	314	0	0	0	0	0
0	0	0	105	609	314	0	0	0	0
0	0	0	0	100	589	315	0	0	0
0	0	0	0	0	91	585	321	0	0
0	0	0	0	0	0	112	579	319	0
0	0	0	0	0	0	0	108	573	314

$\chi^2 = 34686$

into a 10×10 array. A typical example of a partitioning of 10,000 pairs is shown in Table A.3, where the pairs of prn's are clearly not uniformly distributed. The value for χ_T^2 associated with 1000 repetitions of the test was 99,999.

The generator (A.14) fares no better with the runs test. As shown in Table A.4, the distribution of run-ups of length r does not resemble the expected distribution given by (A.11). Thus the results of the three tests shown here would encourage us not to use generator (A.14) in our Monte Carlo calculations.

TABLE A.4. Observed Number of Runs of Length r Compared with the Expected Number after 10,000 Samples for Pseudorandom Number Generator (A.14)

Length of run	Observed	Expected[a]
1	7664	1667
2	1168	2083
3	0	916
4	0	263
5	0	58
6 and over	0	12
	$V = 28370$	

[a]Calculated using Eq. (A.11).

A.3. PSEUDORANDOM NUMBER GENERATION ON PARALLEL COMPUTERS

The advent of parallel processing on modern computers poses new and distinctive problems for prn generators. They may be summarized as follows. Since parallel computers and other supercomputers will be capable of very large calculations, very long pseudorandom sequences will have to be available. Second, it remains highly desirable that individual runs be reproducible. The new difficulty is that on multiple-instruction-stream–multiple-data-stream (MIMD) machines neither the order in which operations are done nor the identity of the processor that carries out a specific portion is necessarily the same when a calculation is repeated. Neither a centralized prn generator nor even a separate one for each physical processor will guarantee reproducibility. Finally, it will be necessary to ensure the effective independence of different realizations of a Monte Carlo calculation carried out on the different processors of a parallel machine. Since architectures have been proposed with thousands of processors, it is a challenge to provide thousands of sequences that predictably reduce the variance by the number of distinct realizations of the calculation.

The first of these problems can be solved by the use of sequences with very long periods. Tausworthe generators are suitable here. Unfortunately, they are difficult to initialize. It is not known how to provide many independent sequences. Long sequences may also be obtained from the composite generators proposed by Marsaglia.[14] Note that if g parallel Monte Carlo computations are being carried out simultaneously and if one assigns each processor in advance to use an equal portion of the original sequence, then the period for each is reduced to τ/g.

Reproducibility of Monte Carlo runs is, in principle, easy to ensure. It is simply necessary that each initialization of a random walk and each "spawning" of a daughter walker (as in a branching or multiplying system) be given a fresh, independent prn sequence. This was first pointed out by Zhong and Kalos[16] and leads to the concept of a pseudorandom tree, in which one generator is used for "intraprocess" random numbers, another for initialization.

A concrete suggestion along these lines was made by Fredrickson et al.[17] in which congruential generators are used for both. They call this a *Lehmer tree*. Starting with x_0, the root of the tree, every element x has exactly two successors, a left successor x_L and a right successor x_R:

$$x_L \equiv (\lambda_L x + b_L)(\text{mod } P),$$

$$x_R \equiv (\lambda_R x + b_R)(\text{mod } P).$$

The constants λ_L, λ_R, b_L, b_R, and P determine the generator. The intraprocess sequence might be that obtained by using the right successors. The left successor is used only to create new sequences, parallel to the first. Thus, the constants λ_R, b_R, and P must be chosen to satisfy the criteria for a good multiplicative congruential generator. The authors proved a necessary condition for nonoverlapping of the resulting sequences.

One might also consider the possibility of using a Tausworthe generator for creating seeds. Unfortunately, it requires that much information—the entire list of saved numbers that specifies the "state" of the generator—be copied from one computational process to another.

The question of independence of separate sequences to be used in parallel remains a major research issue. Not enough is known about the long-term correlations within linear congruential generators to use equal subsequences with confidence.

REFERENCES

1. It may seem that we are glossing over the fact that prn's are not truly random. However, see M. Kac, What is random?, *Am. Sci.*, **71**, 405, 1983.
2. D. E. Knuth, *The Art of Computer Programming*, Vol. 2.: Semi-numerical Algorithms, Addison-Wesley, Reading, Massachusetts, 1969.
3. H. Niederreiter, Quasi–Monte Carlo methods and pseudo-random numbers, *Bull. Am. Math. Soc.*, **84**, 957, 1978.
4. E. J. Dudewicz and T. G. Ralley, *The Handbook of Random Number Generation and Testing with TESTRAND Computer Code*, American Sciences Press, Columbus, Ohio, 1981.
5. J. von Neumann, Various techniques used in connection with random digits, NBS Appl. Math. Series #12, U.S. Government Printing Office, Washington, D.C., 1951, pp. 36–38.
6. D. H. Lehmer, Mathematical methods in large-scale computing units, *Ann. Comp. Lab. Harvard Univ.*, **26**, 141, 1951.
7. B. Jansson, *Random Number Generators*, Almquist and Wiksell, Stockholm, 1966.
8. H. Niederreiter, Pseudo-random numbers and optimal coefficients, *Adv. Math.* **26**, 99, 1977.
9. G. S. Fishman and L. R. Moore, An exhaustive analysis of multiplicative congruential random number generators with modulus $2^{31}-1$, *SIAM J. Sci. Stat. Comput.*, **7**, 24, 1986.
10. G. Marsaglia, Random numbers fall mainly in the planes, *Proc. Nat. Acad. Sci. USA*, **61**, 25, 1968.

11. R. C. Tausworthe, Random numbers generated by linear recurrence modulo two, *Math. Comp.* **19**, 201, 1965.
12. H. Niederreiter, Statistical tests for Tausworthe pseudo-random numbers, *Probability and Statistical Inference*, W. Grossman et al., Eds., Reidel, Dordrecht, 1982.
13. S. Kirkpatrick and E. P. Stoll, A very fast shift-register sequence random number generator, *J. Comp. Phys.*, **40**, 517, 1981.
14. G. Marsaglia, A current view of random number generators, Keynote Address, Computer Science and Statistics, XVI Symposium on the Interface, Atlanta, Georgia, 1984.
15. I. Borosh and H. Niederreiter, Optimal multipliers for pseudo-random number generation by the linear congruential method, *BIT*, **23**, 65, 1983.
16. Y.-Q. Zhong and M. H. Kalos, Monte Carlo Transport Calculations on an Ultracomputer, Ultracomputer Note #46, CIMS, New York University, 251 Mercer Street, New York, New York 10012.
17. P. Frederickson, R. Hiromoto, T. L. Jordan, B. Smith, and T. Warnoek, Pseudo-random trees in Monte Carlo, *Parallel Computing*, **1**, 175, 1984.

INDEX

Antithetic variates, 92, 109–111, 112
  examples, 110, 111
  with importance sampling, 111

Bias, 4, 35–37, 84, 113, 167
  definition of, 35
Binomial distribution, 13, 14
Bivariate distribution, 21–23, 47, 48
Bloch equation, 157
Boltzmann equation:
  distribution function, 119, 122, 125
  integral equation, 148, 156
Bosons, simulation, 123–125, 157
Box–Müller sampling technique, 47, 48, 53, 70
Branching process, 154, 155, 179
Brown, F. B., 70, 71

Cauchy distribution, 20, 21, 27, 30, 31
Central limit theorem, 26, 27, 30, 48, 99, 159
  to sample Gaussian, 48
Central moments, 11, 27
Characteristic function, 43, 44
  definition of, 30, 31
Chebychev inequality, 26, 27
Classical simulations, 118–122
  hard sphere potential, 118–122
  Lennard–Jones potential, 118, 119
Composite events, 7, 8
Composite prn generators, 172, 179
Composition techniques, 52–61, 70, 99
  Marsaglia's method, 70
Compton scattering:
  law, 143
  wavelength, 142
Comte de Buffon, G., 4
Conditional:
  expectation, 23
  probability, 22, 23, 64, 74, 137
Control variates, 92, 107–109
Convolution, 30
Correlation coefficient, 5
  definition of, 13
Courant, R., 5
Covariance, 12, 13
Cumulative distribution function, definition of, 16–18

Debugging codes, 34, 86, 87, 169, 170
Dirac delta function, 16, 143
Distributions:
  continuous, 15–21
    sampling, 39–49, 54–70
  discrete, 13–15, 16, 18, 32
    sampling, 39, 50–53
    for vector computer, 70, 71
  mixed, 39
    sampling, 52, 53
  singular, sampling, 44, 66, 73, 97–102
Divonne, Monte Carlo integration code, 115

Efficiency of Monte Carlo integration, 91, 92
Elementary events, definition of, 7–9
Ergodic, 76, 77
Erpenbeck, J. J., 73

Estimator, 26, 27, 35–37, 89
  consistent, 36
  definition of, 24, 25
  for variance, 28
Expectation value:
  bivariate distribution, 22
  conditional, 23
  continuous case, 18–21
  discrete case, 10–12
  in Monte Carlo quadrature, 24, 25
Expected values, 11, 103–107
  in $M(RT)^2$, 105–107
Exponential distribution, 20, 30
  sampling, 44, 45, 110

Feller, W., 76
Fermi, E., 5
Fermions, simulation, 126, 127, 157
Fishman, G. S., 171
Fourier transform, 30, 121
Fredrickson, P., 179
Friedrichs, K. O., 5

Games of chance, 1, 89
Gaussian distribution, sampling:
  Box–Müller method, 47, 48
  Kahn rejection, 68–70
  table look-up, 70, 71
  see also Normal distribution
Geometric distribution, 14, 15
  sampling, 52
Gossett, W. S., 5
Green's function, 157, 160, 161, 163, 164
Green's function Monte Carlo (GFMC), 127, 157–168
  importance sampling, 165–167
  Neumann series, 159–161
  one-dimensional example, 161, 162, 163, 166
  and random walks, 162–165
Green's theorem, 125

Importance sampling:
  examples, 94–96
  in GFMC, 165–167
  integral equations, 149–151
  optimizations, 96, 97
  radiation transport, 99–102
  singular integrand, 97–99
  and zero variance, 93, 94, 166

Independent events, 8, 12, 13
Integral equations, 145, 147–166
  Boltzmann equation, 148
  branching process, 154
  expectation value, 147, 148, 150, 151
  importance sampling, 149–156
  and random walks, 147, 162–165
  zero variance, 152, 153, 166
  see also Green's function Monte Carlo (GFMC)
Integration, Monte Carlo, 23–25, 31–37, 92–115
  central idea, 25
  efficiency, 91, 92
  error, 89, 90
  vs. numerical quadrature, 90, 91, 115, 116
  see also Variance reduction
Ising model, 122, 123

Joint probability:
  continuous, 21, 22, 28, 63, 64
  discrete, 8, 9

Kahn, H., rejection technique for Gaussian, 68–70
Kalos, M. H., 179
Kinetics, Monte Carlo, 74

Laplace, P., 4
Law of large numbers, 25, 36
Lehmer, D. H., 170
Lewy, H., 5
Lord Kelvin, 4

Marginal probability:
  continuous, 22, 64, 71, 103–105
  discrete, 9, 70
Marsaglia, G., 171
  combination generators, 172, 179
  composition method, 70
Mean value, 10, 11, 18–20, 24–27, 35–37, 158
Metropolis, N., 5, 73, 78, 105, 117. *See also* $M(RT)^2$ algorithm
Molecular dynamics, 122
Monte Carlo:
  central idea, 25
  evaluation of integral, 32, 33, 89–116
  evaluation of sum, 31, 32

fundamental theorem, 25–28
history, 4, 5
Moore, L. R., 171
$M(RT)^2$ algorithm:
acceptance probability, 74, 75, 78
applications, 84, 85, 117, 119–121, 123, 124
asymptotic distribution, 75–78, 82, 83
detailed balance, 74, 77
elementary description, 74–78
example, 80–83
kinetics of, 74, 85
usage, 78–83, 85, 86
use of expected values, 105–107
Multiplicative congruential generators, 170, 171
Mutually exclusive events, 8

Negative correlation, 13, 92, 109. *See also* Antithetic variates
Neumann series, 159–161
Neutron transport, *see* Radiation transport
Niederreiter, H., 171, 173
Normal distribution, 19, 20, 43, 44
central limit theorem, 26, 27
characteristic function, 30, 31
sampling, 47, 48, 68–71
see also Gaussian distribution, sampling
Normalization, 19, 21
rejection techniques, 61, 64

Poisson distribution, 15
Power law, sampling, 44, 55–57, 60
Probability, definition of, 8, 9
Probability density function (pdf):
definition of, 16–19
examples, 20
Probability distribution function (pdf):
definition of, 16–19
examples, 20
Pseudorandom number generation:
combination generators, 172
mid square method, 170
multiplicative congruential generators, 170, 171
on parallel computers, 179, 180
Tausworthe generators, 171, 172
Pseudorandom numbers (prn), 33, 34, 40, 169
generation, 170–172
on parallel computers, 179, 180
testing, 87, 172–178

Quadrature, Monte Carlo, *see* Integration, Monte Carlo
Quantum systems:
bosons, 124, 125, 157
fermions, 126, 127, 157
GFMC, 157–167
Queuing systems, 109, 129

Radial distribution function, 79, 121, 125
Radiation transport, 130–141, 145, 146, 148
Boltzmann equation, 148, 156
example, 131–141, 153–156
importance sampling, 100–102, 145, 153–156
neutral radiation, 130
Random:
events, 2, 7
numbers, 2, 34, 169
sampling, 1–3, 39, 40
Random variable, uniform:
definition of, 32, 39, 40
pdf, 44, 45
see also Pseudo-random numbers (prn)
Random walks, 5, 145–148, 157, 162, 163
expectation values, 147
in $M(RT)^2$, 75–79
Rejection techniques:
description, 61–65
distribution function, 63–64
efficiency, 61, 64–65
examples, 62, 63, 65–70
singular pdf, 66
Rosenbluth, A. W. and M. N., 73

Sampling:
angle:
uniformly, 48
von Neumann method, 66–68
definition of, 39–40
sum, 59–61
testing code, 86, 87
Schrödinger equation, 123, 124, 157, 160, 161
Score, 151–155
Simulation, 2, 3, 109
classical system, 117–123

Simulation (*Continued*)
quantum system, 123–127
radiation transport, 129–141, 145, 146
solitaire, 2, 3, 32
Standard deviation, 11, 21
estimator, 28
Standard error, 11, 27
Statistical mechanics, 5, 74
in physics, 117–127
Stochastic process, 2, 7, 15, 39
examples, 130, 132, 145
Stratification methods, 112–115
in many dimensions, 113–115
variable interval size, 113
Student, *see* Gossett, W. S.

Tausworthe, R. C., 171
prn generator, 171, 172, 179, 180
Teller, A. H. and E., 73
Tests of prn generators:
equidistribution test, 172, 173
example, 174–178
runs-up and runs-down, 173, 174
serial test, 173
Thomsen cross section, 143
Torrie, G. M., 73

Ulam, S., 2, 5

Valleau, J. P., 73
Variance:
continuous, 19, 22, 24
discrete, 11, 12
estimator, 28
Variance reduction:
antithetic variates, 92, 109–112
control variates, 92, 107–109
expected values, 92, 103–107
importance sampling, 92–103
stratification methods, 112–115
Variational calculations, 124, 125
von Neumann, J., 5
mid square method, 170
rejection technique, 66–68, 142

Whittington, S. G., 73
Wiener, N., 48, 86
Wood, W. W., 73

Zero variance calculations, 93, 94, 152, 153, 166
Zhong, Y.-Q., 179